ALLAN BLACK

HOW TO BUY A CAR

A BUING-GUIDE FOR DUMMIES

IMPRINT

ISBN 978-1090394705

Printed and bound by KDP - Kindle Direct Publishing

Published by O. Sharanych Media[SM]
160 00 Praha 6, Czech Republic

FOREWORD

Every day a growing number of people wish to become the proud owner of their own car. This isn't surprising as the advantages that a personal car provides are far beyond the care and costs associated with it.

Most people buy cars on the market through advertisement over the internet or through classifieds (in other words, they buy the car from „an unknown person"). More affluent people buy a car in a Showroom. However, in any case each and every buyer has to face the problem of choice. And the selection process can be fraught with many surprises, including unpleasant ones.

This book is for the people who intend to buy a car.

So, how not to be deceived when choosing and buying a car? How to avoid becoming a victim of fraudsters? How to make your memories of buying a car a pleasant one? What are the things that require special attention when choosing and purchasing a car? This book will guide you along with recommendations, useful tips, extensive knowledge and interesting information.

What are the primary factors that should be considered when buying a car? How to start a car inspection? Why is it not recommended to go to the market alone and inexperienced? Answer to these questions and as well as many other things, will be covered in this book.

TABLE OF CONTENTS

PREHISTORY

The story began over 120 years ago. It was created by German researchers Gottlieb Daimler and Karl Benz, the first internal combustion engine. It was a breakthrough in the creation of the first unit. Since that time, began the era of engineering.

The first inventions of devices on the steam engine began to appear in the 17th century. They were like crews. They don't move quickly, made loud noise and also produced a lot of smoke.

It is considered to be - 1885, the year of the creation of the first gasoline engine. The name then he received „Motorwagen". It was invented by two German engineers, Benz and Daimler. Three years later, „Benz" released the first four-wheeled cars.

For the record, the first truck equipped with an internal combustion engine and a cargo battery appeared in 1896. The analogue with a diesel engine saw the light of the world only in 1923.

Today the car is a complex technical machine, consisting of several thousand parts. Every manufacturer makes unique cars in their own way. Their teams of engineers work on all aspects and they approach the same technical tasks from different angles. Therefore, each model has many features, some of which may simply not suit you. Sometimes many firms experiment boldly with new technology and go ahead to manufacture a lot of units, only to identify the multiple flaws in the first few months after they sold the vehicle.

Over the time I have worked on hundreds of cars, variety of brands, models and configurations. Thanks to my experience over the years, I can consider myself an expert in the automotive business. My friends and acquaintances who want to buy a good used car regularly seek my advice on how to do it right. And each time I have to share the same information with them, which is quite a lot. This prompted me to create a step-by-step instruction Manual so that when choosing a car, you wouldn't lose your savings. The reader's potential benefit from reading this instruction is in the scale of TENS THOUSANDS of EUROS. It's theses small and insignificant mistakes that people make when buying a used car that can result into a huge loss as they later become the victim of dishonest sellers.

From experience I could say that, among people who buy used cars, even the people with lots of wealth and experience don't do many things that are need to be done. And at the same time they make a lot of avoidable errors due to overthinking the situation.

If you are far from the automotive field, this guide is an invaluable source of information for you!

I have outlined the process of buying a car in simple and understandable language with photos. My goal is to save you time and money, and my instructions are based on this.

Finding a good used car isn't easy and anyone who has experience buying a new car at least once, knows it.

LET'S START STEP BY STEP

WHAT IS THE PRIMARY PURPOSE?

Example, you may need a car for daily trips to work, trips to shops and the market, or take your whole family along to the countryside to enjoy the nature etc. You might have many other purposes in mind apart from these. To make the right choice from a variety of options, you ask yourself what you intend to transport. The right answers to these questions will tell you what kind of car you need. Many people tend to first think about the things they need to transport.

If you intend to drive a car to work, then a middle-class car with a sedan-type body will suit you. For example, Opel Vectra, Toyota Corolla, etc. Some Compact-class cars are also well suited for this purpose. With body type „hatchback", „coupe" or „sedan": „Opel-Astra", „Volkswagen-Golf", etc. (Fig. 1.1).

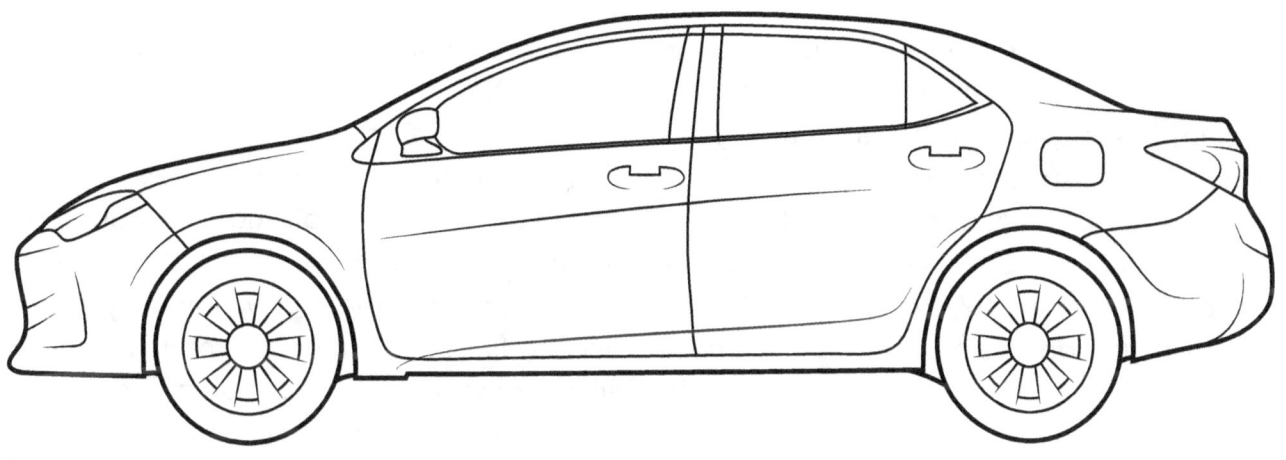

Fig. 1.1

For family trips outside the city, it is convenient to use roomy cars with the "universal" body type: Ford Mondeo, Opel-Omega, Mercedes, etc. Minivans are even better for such purposes. True, minivans are in a different price category and cost a lot, but for a large family, it's hard to find a more convenient vehicle.

For trips to shops and markets (i.e. for shopping), women are happy to use small-sized cars (Fig. 1.2), such as the Ford-Ka or Skoda Fabia.

Fig. 1.2

These cars are easy to drive. Due to their small size they maneuver excellently through the city and at the same time they have enough space to accommodate your purchases. It is obvious that it's not a good idea to go to the supermarket on a „station wagon" or a „minivan" (unless you plan to stock up on food and other things until next year).

Step 1. Apparently the choice of a particular brand and model of the car is not all. Before buying, you need to carefully assess your financial limits and at least approximately think about how intensively you intend to operate the car. based on this you need to formulate your requirements to opt for the year of production and the general technical condition of the car. Simultaneously, it is recommended to determine the color (if this is of fundamental importance), as well as various kinds of "Accessories" (stereo,

electric equipment, air conditioning, etc.) you want.

Step 2. If you intend to buy a new car, then the second step is a bit more complicated. It depends on your choice, which will either be absent altogether or it will be very superficial to find. For example, "you may want a silver color car with air conditioning". In the case of used cars, it's a wide field of search. Now, this leads you to a couple of questions such as: Whether it is critical for you if there are a couple of "Dents or scratches" on the car body (i.e., places where rust begins), or you want the body to be in excellent condition? If the car is good in all aspects with minor or just one drawback (Eg: A worn tire that needs to be changed), will it suit you or not? Various such questions can be asked, which makes the second stage of the process a one which needs thorough thinking.

Step 3. At this stage, when you have decided on your requirements for your future pleasure car, you choose a particular car in the market. And here it is important to know and understand that a car is never bought on the first sight. This applies primarily to used cars. Importantly you need to take a closer look, study the prevailing market trends, etc.

WHAT ARE THE CLASSES OF PAS-SENGER CARS?

Just before a hundred years ago, at a time when there were a very few cars on the street, classification of cars was very simple, namely Cars and trucks. And based on the fuel it was either gasoline or diesel.

Things changed when the hard efforts of many engineers who were trying to improve the body and the components of passenger cars started giving results. More and more new models began to appear in the markets of different countries, with various sizes, appearance and technical characteristics. In order to somehow streamline this abundant set of differences, classification of cars was introduced, dividing them into several categories.

In today's world cars are divided according to the American, European and Asian classification systems. Russia follows the European version. The European classification includes 10 group of cars, which has 6 size classes and 4 body class. Let's get closer to how you can determine the class of the car by their external appearance.

CLASS A -SUPERMINI

Classification of cars by class includes in this category compact cars designed for use in urban streets. Usually they do not exceed 360 cm in length and 160 cm in width.

Thanks to their size, models representing group A can be easily recognized amongst a huge group of cars. Whatever may be the logos on these cars, if you see a compact three-door hatchback in front of you, the dimensions of which correspond to the above figures, you can be sure that you met a car from class A. 5-door hatchbacks are also included in this segment occasionally.

The main advantages of these mini-cars are high maneuverability and compactness. They are not able to exhibit great speed, but speed is not required on city streets. The cabin accommodates the driver and passenger and two more passengers sit in the back, but there isn't much room for comfort at the back. Europeans, who prefer economical classes of cars are very much interested in these nimble vehicles and buy them as a second car in their garage. Typical examples: Renault Twingo, Citroen C2, VW Lupo, Daewoo Matiz.

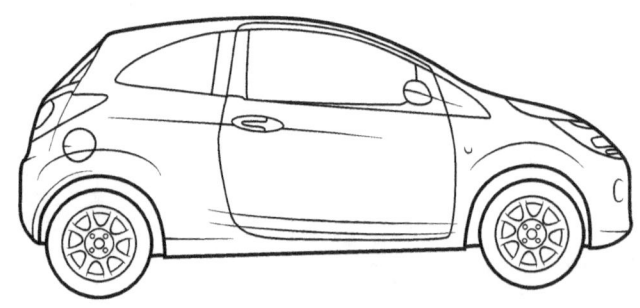

CLASS B - SMALL

According to the European classification, the group B class includes the vehicles whose body length does not exceed the mark of 390 cm, and the width not more than 170 cm. This category is also referred to as subcompact. They are often found on the streets of Mediterranean cities. Most models are front-wheel drive hatchbacks with 3 or 5 doors and occasionally you can see a sedan.

In terms of equipment and comfort, they are better than class A cars but they are inferior to more advanced Class C models. Thanks to their small size, cars from this category are in high demand amongst female drivers. The cabin is quite spacious, but the rear seat is designed for two adult passengers rather than three. Typical Examples: Ford Fusion, Hyundai Getz, Fiat Punto.

CLASS C - MEDIUM SMALL

Middle class cars account for the maximum share of sales in the European market. The commonly adopted classification system refers to cars from segment C as golf class, which was derived from the brand name that introduced and marketed this Class for the first time. The car is defined as a class C when their length fits in the frame of 390-440 cm and width of 160-175 cm. This could include cars with any type of body, be it a wagon, hatchback, UPV or sedan.

The cabin of the golf class car is quite spacious and comfortable for five people to travel. The car is designed for long distance driving and is able to cover significantly long trips through the streets inside the city or intercity routes. Typical Examples: Audi A3, Opel Astra, Toyota Corolla.

CLASS D - MEDIUM OR FAMILY

It's quite simple to determine whether if a car belongs to class D, just by looking at its appearance. These are spacious cars with an attractive design and body dimensions ranging from 460 cm in length and 180 cm in width. Body type could be anything but the primary aspect is a fairly high level of comfort and equipment. The perfect combination of technical characteristics and affordable prices make this segment of cars one of the best-sellers.

There is also a sub division of cars in this class into family and elite. This additional classification allows you to choose the right car configuration, whether if you are a businessman or a housewife, and based on your requirements. Some elite models can compete with sports models. Typical Examples of the family category of class D: Citroen C5, Opel Vectra, Nissan Primera. The elite models are Audi A4, Jaguar X-type, Volvo S60.

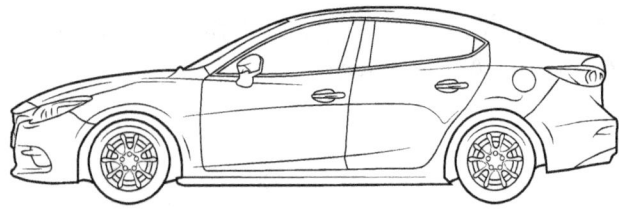

CLASS E - FULL SIZE AND EXECUTIVE

The European classification refers to these cars as the Full size or business class. In Russia, such cars are divided into two segments namely Executive and Higher Medium. In addition to being a sedan and station wagon, these cars can also have a hatchback body. The length of the machine from the group E exceeds 460 cm, and the width is from 170 cm.

The spacious passenger compartment and a high level of basic equipment allow you to drive comfortably for several hours. This is complemented by a large wheelbase and independent suspension (for most models). Typical Examples: Mercedes E-Class, BMW 5-Series, Toyota Camry, Volvo S80 / V70.

CLASS F - LUXURY

The European system classifies cars with a spacious comfortable cabin catering to the highest business classes in this segment. Cars of the „luxury" segment are the highest class due to the level of equipment used and the use of expensive materials in the interior. Under the hood of a luxury sedan you can find a powerful 6-cylinder engine providing excellent dynamic performance, and the large dimensions (length - from 460 cm, width - from 170 cm) allow passengers to feel as comfortable as possible.

Many models have a pronounced sporty appearance. These cars are usually driven by a chauffer, which exhibits the status of its owner. The luxury segment includes models such as the Jaguar XJ8, BMW 7-Series, Audi A8, Rolls-Royce Phantom.

In addition to the above, this classification also includes car body segments.

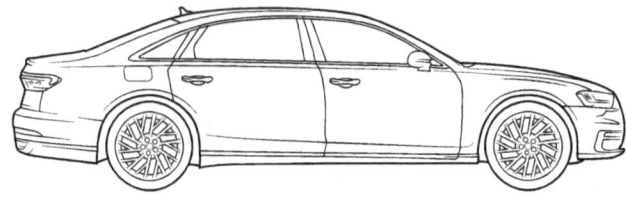

S - COUPE, CABRIOLETS, ROADSTERS

These cars have the appearance and character of sports models. They are usually designed for two or sometimes four passengers. They are distinguished by their low slung and stiff suspension. Few examples are Mercedes-Benz SLK, Audi TT Coupe, Porsche 911.

M - MINIVANS, MPV

This segment is characterized by their capacity, accommodating a maximum of eight passenger seats (without driver). If the number of seats exceeds this number, then they are called minibuses (segment M1). Examples are the following cars: Honda Odyssey, Nissan Quest, Mazda 5, Opel Zafira, Ford C-Max.

J - SUVS, CROSSOVERS

This includes all-wheel drive passenger cars and utility vehicles that are used for active recreation, professional and as general purpose off-road vehicles. Typical examples are the Mercedes G wagon, Hummer H1, Subaru Legacy, Ford Expedition.

These also include crossovers. These cars are characterized by a combination of characteristics of several different classes. They represent an intermediate link between the wagon and SUV. This class has a few examples such as: Infiniti FX, Lexus RX300, Nissan Murano.

Here are a few examples: Fiat Strada, Ford Ranger, Dodge Ram, Nissan Titan, Volkswagen Amarok.

Pickups, mini pickups, pickups, giants

These are Vehicles for the transport of various goods. Most often they are all-wheel drive or rear-wheel drive cars.

CAR BODY

The type of car body is a specification found in the descriptions of all passenger vehicles. It depends on a number of features such as the operation of the car, utility of the car, its cargo capacity and many other parameters.

The type of car body is defined by the factor that characterizes the ratio of the internal volume of the body to its cargo capacity, the technical features of the luggage, bonnet compartments and layout of the passenger compartment. Simply put, the capacity inside as well as the size and configuration of the luggage compartment, depends on the body type.

The presence of a large number of body types is due to the fact that it is extremely difficult to structurally combine the cargo capacity and design features, which forces developers to create various different variations designed to satisfy people with different requirements. Example, for people those who regularly transport a large number of goods to people for whom cargo capacity is of the least importance and for those whom the design and driving characteristics of the car matter.

CLASSIFICATION OF PASSENGER CARS BY BODY TYPE

Today there is a huge number of different types of car bodies. Automakers tend not to stop with the existing ones and they experiment inventing new types and trying to draw attention to their new products with creative advertising about their „innovation" of a new body type. Such experiments can include the recently appeared category called "four-door coupe", "off-road coupe" and others. However, in our review we will try to stick to the traditional options, since all the above mentioned versions are variations based on them.

So a SEDAN is car in a three-volume body and is the most well-known version in our country. According to statistics, in Russia this type of bodywork is the most popular, which is probably because people are used to such kind of cars as they have been around for a long time now. It is connected with the fact that most of the Soviet cars were exactly sedans and this standard shape is well known amongst people. Conventionally speaking, a sedan is a car in which the engine compartment, the passenger compartment and the trunk are separated from each other and have separate access. Today, this type of body is very popular although it has undergone certain refinements over time. Also speaking, the luggage compartment of sedans can no longer said to be completely independent in modern cars.

They are just separated from the passenger compartment by the back of the rear seat, which, if necessary, reclines, allowing you to increase the trunk capacity.

COUPE

The Coupe is a body type that appeared in the early twentieth century. It is more usually seen in sports cars with a three volume body fitted with just a pair of doors having increased dimensions which makes it easier for passengers to enter and sit in the back seat. And the rear seats are often absent in many sport variants. Also, there exists a special class of premium coupes, which are designed based on sedans, intended for wealthy customers who prefer to drive their own car.

HATCHBACK

'Hatchback' is the most popular type in Europe due to its utility coupled with overall compactness. It is a two-volume body, where part of the passenger compartment is assigned to the trunk. The design of the fifth door which is the Tailgate, is flat and opens either upward or outward giving good dynamism to the car.

The advantage of the hatchback is the versatility to transform the load compartment and the cabin which gives room to more space and increased capacity as well as the convenience of carrying long loads. You can see that such a layout allows you to reduce the size of the car while maintaining capacity, which is an important advantage for urban cars.

LIFTBACK

Liftback is considered to be a specific kind of hatchback. The difference is the configuration of the fifth door, which opens up like a sedan and is projected outward lifting up with the rear window. It's common for the for a manufacturer to use a double opening, the trunk lid or the entire door can be opened, making the car closer to sedans.

The advantage of a Liftback is greater than those of a hatchback as they could offer spaciousness while maintaining the traditional proportions of a sedan. Many brands use a similar layout for business class cars, where the traditional shape is an important element of prestige.

STATION WAGON

The body type of the station wagon, as the name suggests is designed for maximum functionality. Construction wise, it is similar to a hatchback with the only difference being that the luggage compartment is much more spacious due to the vertical position of the fifth door, and by maintaining a high roofline throughout the body. Today, wagon-type cars are more often associated with utilitarian cars, but in the mid-twentieth century a 'shootingbrake' body configuration gained popularity which is a kind of extended universal body with a three-door configuration that combines the dynamics of a classic coupe with the advantages of a universal layout.

Today, this type of bodywork is being revived by Mercedes Benz in some models with the word "wagon" prefixed to name of the vehicle. However, such station wagons differ from the classic models by the presence of a five-door layout and with a sporty body.

MINIVAN

A minivan or as it is sometimes called 'A Monocab', is a type of bodywork in which the engine compartment, the passenger compartment and the trunk are arranged in a single box or body. This body is much higher rather than that of the wagon, which allows for wider possibilities to transform the cabin. Such cars usually have a seven seater layout without increasing the overall dimensions by implementing vertical rows of seats back to back.

To be specific this class of car is a cross between a station wagon and a minibus, combining the convenience of both body types.

PICKUP

The United States of America is known as the homeland of pickups, where these cars are very popular. The pickup is a SUV with a separate open cargo compartment which resembles a kind of mini-truck for the transportation of cargo. This car gained popularity amongst people in the agriculture related sectors due to their spaciousness, carrying capacity and as the luggage compartment could easily be cleaned with the help of available tools. In modern times, pickups have shed their functional image and are often sold as a car for outdoor activities. This is not just because of the space and room they offer, but also because of their capability to maneuver better in off-road terrain conditions and the durability of their suspension.

Today the comfort offered in pickups are as good as or similar to SUVs.

OFF-ROAD VEHICLE/SPORT UTILITY VEHICLE (SUV)

The SUV is a car with a universal body type, with an all-wheel drivetrain and a body on frame structure. They allow you to overcome difficult road conditions and have high ground clearance. In recent years, the SUV market has been divided between Classic frame models and crossovers which are universal cars with increased ground clearance with a Monocoque body.

Most often the basic variants of SUV's do not have all-wheel drive or a serious arsenal of off-road capabilities, but they are versatile and roomy. Today, crossovers have practically replaced traditional minivans, which actually over generations actually turned into crossovers and got a similar layout.

CONVERTIBLE/CABRIOLET, ROADSTER

A convertible is a car with a removable fabric or metal top, most often with two-door layout. There were four door models in the past but however due to difficulties in maintaining the rigidity of the open body, they didn't become widespread. The most common version of the cabriolet is a roadster which is an open top sports car with a two-seater cabin or a 2 + 2 seat arrangement, i.e. a rear seat is provided as an extra luggage compartment or a child's seat.

Predominantly in the case of roadsters, manufacturers provide them with sport controllability and brand them as an alternative to the sport coupe. Almost all sport car manufacturers have such models available in their product lineups.

LIMOUSINE

The body type „limousine" is not independent but a variation of the classic sedan (and often an SUV). Limousines are usually produced by third-party tuning companies by inserting an extra body frame into the central part of the car increasing the length of the cabin in the required range (from 30 cm to several meters). The primary aim of the modification is to increase the capacity of the rear cabin to accommodate a rich interior, additional technical, entertainment features and to provide more space.

Usually limousines are used for the transport of dignitaries and are equipped with lots of safety features. Mercedes Benz is the only automaker that manufactures factory produced limousine with its model Pullman. In addition to that, most companies that manufacture executive class sedans usually produce them in two wheelbase options namely, standard and increased. There is a huge market for modifying conventional sedans and SUV's into rental Limousines. On the interior, they are significantly different from the conventional sedans. They are roomy rides on wheels with a mini-bar, audio system and comfortable sofas inside.

VAN

As early as the end of the twentieth century the van was a separate class of cars where, there was a separation of the driver cabin (usually double) and the closed cargo compartment by a door and they had no windows. Such cars were a convenient means of urban transport for delivery of goods.

Today, manufacturers have moved away from the conventional approach and usually offer a standard station wagon or minivan with a few row of rear seats, a large cargo area and missing windows (a good example of this is the modification of the Lada Largus). Such an arrangement allows access to the cargo not only just through the rear loading door, but also through the standard side doors of the car. From the passenger compartment the cargo compartment of the van is separated by a metal partition.

You should consider a number of factors, before coming to a conclusion about the choice of particular type of carbody you are going to choose. Out of these, the main ones are the dimensions of the desired car and your needs regarding the carriage of passengers and cargo. So we saw that Hatchbacks and station wagons offer the ideal balance between capacity and comfort, while the ones with smallest capacity are sports coupes and roadsters. Buying a pickup truck or a van is preferable in cases where the transportation of goods is a priority. In the case of a large family car minivans as well as crossovers exhibit the best convenience and comfort for passengers. Traditional sedans are an average balance in terms of cargo capacity and passenger comfort. Summarizing it all, it is worth noting that today's market literally offers body types for every need and requirement. Choosing a suitable option, you will get the car that is best suited for your current needs.

The body of a modern car is a product of scientific research and engineering. Every detail and curve has its own purpose. And just aesthetics and convenience are not just the final objectives of designers.

If you plan to use the car for just commuting with one or a few passengers, then the compact sedan will be a very best option. The smaller the car the easier it is to find a parking space and maneuver through tight city traffic.

But if you need a car for your regular leisure trip to a cottage in the countryside then it's a different case. You might need to accommodate kids, family and relatives which requires a good passenger space. And also you might want to transport goods such as seedlings during the spring and the Crop during the autumn. The right cars for this purpose would be a Large rear doored hatchback or a station wagon, they are spacious enough to carry a medium sized refrigerator.

Of course these cars are a bit more expensive than sedans but they are worth every penny for the purpose they serve.

ENGINE POWER

On the contrary to popular belief, engine power has a little effect on fuel consumption in modern cars. Unless you compare it with the same cars with different engine options, it will be almost the same.

For driving inside the city in dense and slow moving traffic, engine power doesn't matter so much. But if you are travelling on a

less populated sub urban highway then the additional horsepower can help in fast and dynamic driving.

Talking about fuel consumption, it is more dependent on the weight of the car, its technical condition, maintenance, load carried and driving style.

EQUIPMENT AND OPTIONS

Comfort and 'extras' are always worth the extra money. Weigh your financial capabilities and consider whether it's too excessive for you to pay for power windows, heated mirrors, air conditioning and other accessories that are far from being necessary all the time. Compare the cost of the car in the minimum configuration and with your desired configuration. Perhaps sometimes it is wiser not to overpay and operate the door mirrors manually by yourself.

The choice of gearbox is quite controversial. Automatic transmission is of course a more convenient choice. Especially when It comes to urban traffic. They are usually expensive than the manual variants so if you can afford to spend some extra bucks on an automatic it's perfectly good! But if you are short in your finances and prefer a manual option it will still drive good.

ESTIMATE THE COST OF OWNING AND OPERATING

Buying a car is just half the battle. Its operation and owning requires lots of systematic expenses.

COSTS DEPENDS ON THE MILEAGE DRIVEN

This includes the expenses incurred for gasoline and lubricants, for periodic maintenance and replacement of parts, for wear and tear of brake pads and wheels, tires, wipers, and so on as you drive the car.

REGULAR COSTS

Insurance, Cost of Inspection, maintenance.

REPAIRS

Modern cars are very reliable and durable. But, like any other machine they may require repair from time to time. Usually the more expensive and more prestigious the car is, the more you have to pay for repairs.

CHOOSING YOUR PERFECT CAR!

Now that you have asked yourself a lot of questions about your transportation needs and have given yourself honest and specific answers, put them all together in one picture and draw the image of your ideal car.

It is wise to add, it's important to stay within the limits of common sense and not to demand the impossible from the car. You cannot foresee all life situations.

EXAMPLE 1

You might require a car for trips to the countryside once or twice a year to rest with your family of four. So, there should be enough space for things in the car. Since you might intend to go to the sea in the summer, air conditioning should be a prerequisite for a comfortable trip. The remaining options are optional. The optimal solution to your requirements is a spacious station wagon with an air conditioner from one of the good brands.

EXAMPLE 2

If you need a vehicle for daily commute around the city alone it's better to look for a compact car with an electric power steering option available. This helps you to easily maneuver through cramped city streets, traffic and makes parking easier.

When you know what your right car should look like and you figured the budget of your purchase, it's easy to choose the right model.

USEFUL TIPS

You will hear advice, recommendations and assessments from many sources. Consider only those that are supported by facts.

Absolute drawbacks are rare.

More often, any feature or characteristic of a car is an advantage in one situation but a disadvantage in another. For example, a small car has a cramped interior but at the same time offers better fuel consumption. So the question is "what matters to you the most?"

Do not be afraid to make a mistake. You are not going to use the same car forever. Your next choice will be wiser.

HOW TO FIND THE PERFECT OPTION

I often hear people saying that it's impossible to buy a used car in good condition. Arguing that the car is not new so there will be problems with it and you will have to constantly spend money on service. This statement in fact is not true, over the years I have seen my clients choosing good options despite the fact that the cars were already over 7 years old.

I can tell you that it's possible to find a car in good condition with a budget of 1000euros! To elaborate this fact, I'll tell you an experience of mine. A young guy aged 18-20 old, who just graduated from a driving school turned to me for help. He needed an inexpensive car with a budget not more than 1000 euros. We searched for a long time and finally found a car that was in really good condition. The year of release was 1997, so it was pretty old but it was really well looked after. My client was very satisfied with the purchase as the car was good and half a year later he planned to purchase a more expensive car. An old Toyota never fails in half a year of driving!

It is not necessary to buy a new car from the Showroom, but it is more important to choose the right one! Using the information from this book, you will be able to choose the best options among used cars.

Happiness is not just when you own a car but it's when the car you buy works well!

I'll provide you with some examples from the buying experiences I had with my clients.

So, let's begin,

I'll tell you a story that happened recently.

Well, grab your popcorn!

Here is how the story goes…

Once I got a call from the customer who asked me if I could come and look at an Audi A6 Quattro 2.5TDi with four-wheel drive. I was so tensed up from what I heard, because my client on the other side of the phone sounded like he decided everything he needed and only just needed me to do an inspection. So coming to the point we agreed on the place and time of inspection. At first glance it was clear to me that the seller simply wants to push our AUDI dreamer to buy his car. And he almost succeeded, but fate decided otherwise!

So continuing,

Even without a thickness gauge, it was clear that the car was involved in an accident before, but the seller (as in most cases) argued that there were no accidents and the car was only partially dented. I conducted a thickness gauge inspection and there was an evident dent (photo 1), to which the seller replied: „Come on, in our time there are no cars without crashes!" (we previously recall him saying the car was never in an accident before! :)

Going ahead with the story,

Since it was evident the car was in an accident (which was found out using the thickness gauge), logically it is necessary to check the airbags, but the seller assures that they are perfectly fine and after a few minutes of discussion he allowed to remove the covers (alas, the photo was not allowed to be done) and I found that it hasn't been replaced! In Czech Republic! Yes, it happens around here! But what about the indicator on the dashboard which shows defunct airbags! Well they don't show warnings if you connect the sensor stub. It's obvious!

It was time for a test drive and after all this still the buyer has a bit of hope left. I start to check the automatic transmission (oh, and I almost forgot that the oil hasn't been replaced for a long time now) and the gears are slipping.

As we drive, the steering wheel starts to pull to one side (it is understandable, a serious traffic accident led to the alteration of the body geometry), but even with this issue the seller has found an excuse to not fix it which just made him a poor quality like his car.

After the diagnosis there were many errors and the buyer listened to everything carefully.

After a long conversation with his wife who didn't agree with the car, the buyer decided to abandon the purchase (as he told me). But the next day I got a call from an already familiar voice: "Well, what would you advise me about buying the car? The seller is ready to reduce 7000 euros" and after a long conversation I hope the buyer has finally changed his mind about buying this car.

Let's start to analyze this case:

Firstly, you have found an advertisement/offer on the internet that that you like. What do we do first? Yes, right, we call the seller and ask „Is the advertisement/offer still valid? Where should I come to see?" and quickly go to the meeting place because it could be picked up by someone else before we arrive. Yes, I agree that's possible, but that is not a reason to go at once especially if you don't have much experience in buying a used car. The first thing to do is to check through the database that the car is not listed in stolen vehicles or it isn't leased. Then, you should ask the seller "Where was this car bought? How many previous owners were there? Has the car been in any accident before?". Naturally no one would tell you the whole truth but from listening to the response to these questions one can come to a conclusion whether it is worth going to take a look or not.

Returning back to the case of the Audi, as soon as we arrived at the place to inspect the car, the first and most important thing I began to check is the body. While checking the body one must pay attention to the following points: The car must be clean, otherwise there is no point in inspecting the body. Sit down near the right headlight of the car and carefully inspect the right side of the car to find out whether there are any dents, defects in the surface geometry and color. Repeat this procedure from the left side and then inspect the hood and roof with the same method. From this angle, almost all the 'fixed up/patched up' places are visible. It is advisable to conduct a similar inspection from the trunk. If anything causes a suspicion to your eye use a magnet to check, if it falls off the body it means that there is a filler at that spot. Carefully inspect the joints (between the hood and the fenders, mid-section and doors, etc.) as this is a good indicator in finding out about the past of the car. The gap between the joints have the same width along their entire length. Also pay attention to the stickers and labels which are usually used to mask the defects. Check all the doors and they should close with the same effect and sound. Otherwise it means there is a defect in the body geometry. If you have doubts about the correctness of the body geometry, then you can check it accurately on a special stand designed for it. You can restrict yourself to a simple check and diagnosis for now.

You can identify a repainted or patched up car from the following features: Firstly, there are traces of paint on rubber and the plastic parts of the body, next the color of the paint under the rubber seals (they must have been slightly unscrewed during painting) of the glass, doors, trunk lid. And finally, if you can see the difference in paint texture inside the engine compartment and under the hood. Open the hood and carefully inspect the attachment points of the bumper and side panels. If they have metal folds and chipped paint, the car has been dismantled before. Coming to corrosion each car has its own weak points. But usually this process begins with the fenders, sills and areas under the rubber seals. A proper corrosion check requires the car to be lifted up. When inspecting the bottom of the car feel free to poke it with a screwdriver and you can identify the rusty places better.

HOW TO TREAT ASSIS-TANTS AND ADVISERS WHEN CHOOSING A CAR?

Excessive arrogance and self-confidence is the problem of many buyers. Often people who have just barely started to drive right out of the driving school, having vague knowledge about cars hurry off to buy a car right away without any help. They justify themselves by saying that "I am an adult who is not any stupider than others, so I don't need advisers to confuse me and I'm quite able to buy a car by myself."

There is no surprise that such an opinion is deeply a mistaken one! Only experienced motorists, who not just have years of driving experience but are also well versed in the structure and working of the car could make such a claim to be an expert. If a person has been repairing and servicing cars for many years (for example, he worked at a service station), then of course he can assess the condition of any used car by himself. But if the buyer is familiar with cars through articles from magazines such as "Top Gear" then /she definitely needs an assistant.

However, finding a reliable expert assistant is quite a problem. Of course, if there is someone among your friends, acquaintances or relatives who is a specialist in cars then consider yourself lucky. But if not?

Remember that not every motorist is suitable for the role of an assistant in choosing a car. Unfortunately, there are a lot of idle people who give advice to everyone, although they themselves have a poor understanding of the machine and in most cases they do not have any understanding at all. Such pseudo-advisers could be easily found

at any parking lot or in any garage cooperative. The most unpleasant thing dealing with them is that they give their "advice" with an authoritative look, in a tone that does not tolerate any objections and it's almost impossible to prove their point of view wrong. It is better not to involve such an "expert" as an assistant for choosing a car. If you have no way to find a reliable expert around you, try to check the exact qualifications of the person who offer their services by verifying his advice and recommendations with specialized and well-known automotive journals and manuals for the vehicle.

Remember that it is unlikely for a person who is used to changing his car every 2-3 years to be a good advisor. Because in such a short period of time usually the car does not create any serious problems for the owner, so in reality this person has hardly any significant knowledge and experience with the car. But on the contrary the owner of an old car which he is driving for 7-10 years or more will certainly be able to help you with practical advice while choosing a car. Because over the years he must have had to deal with various problems, so he should be able to provide good practical help.

This is a strange advice, but you should be careful with the advice of professional drivers who drive for their bread and butter for many years. Because as a matter of fact in large enterprises, usually it's not the drivers who are engaged in the repair and maintenance of cars but there are specialists who are specially designated for this job. So the driver may not know the details of the car's maintenance and its diagnostics. As a matter of this fact, it is difficult for them to assess the condition of a used car.

Please note that to absolutely identify all the shortcomings in a used car isn't possible even for an experienced specialist. You can evaluate the external condition of the car, inspect the interior, drive it to the service center and check the underbody, listen to the engine, make a test drive, etc., but it is nearly impossible to look inside the engine or gearbox and access the degree of wear of the various components.

Currently, in the Russian automotive markets you can find specialists who, for a certain amount of money (which is usually calculated as a specified percentage of the value of a car), offer to help you with choosing and inspecting a car. Similar announcements can be found in newspapers where you can also find someone offering to help.

If you don't have anyone to turn to, and your knowledge, experience is not enough (or you have no idea about it), you can really use the services of such assistants. Sorry to say, but the money isn't worth it in some cases, as your savings go down. But in most cases you can try to make a good bargain.

However, you should always remember this is a double edged sword. On the one hand, the help of a specialist who is well versed in cars is a great advantage. As it's no secret that Russian sellers can disguise a teapot as a car and sell it to you as if it's a really great machine. The assistant will look effectively

and draw your attention to invisible flaws and give you advise which can bring down the price of the car that could discourage you from buying a totally decent-looking car on the outside, but completely rotten in the inside. In addition to that only people who are constantly dealing with cars have enough experience to identify signs of some specific issues. These issues are, for example whether it's a stolen vehicle, whether it has problems with customs or if the engine has been killed previously and replaced etc. Is it very important to mention that these are some serious issues which you need to avoid at all costs.

In other words, a good specialist has a special gut feeling or instinct which he has developed over the years from experience to identify problems invisible to the eye, and this skill is very crucial.

But there is a huge downside. Some of these assistants are DOUBLE AGENTS as on one hand they help the buyer to buy the car and on the other they help the sellers to sell their vehicles. While selling the car the assistant will not care much about the actual state of the car being sold and he will recommend this car to the buyer without hesitation because it's good money for him.

But assistants who are constantly working in a particular market don't recommend their clients into buy cars in poor condition and are fair.

Let's look at a case where at first, the seller finds an expert assistant to assist in selling his car for a fee. The assistant assesses the condition of the car and looks for a client who wants to buy a car approximately in the particular class and condition. Then he assists the client in choosing a car and also gets paid for it. As they go around the market and look at the cars put up for sale the assistant would find flaws in the cars the client is interested and would discourage him into buying them. Thus eventually slowly they will get to the car the assistant is promised a reward for. Is there any doubt that in the end, it is this car that the trusting client would have chosen?

Do I need to take money with me?

Most car buyers can be divided into two groups: those who take money with them right away to buy the car and those who go to choose a car without the cash and take the car later after the settlement with the seller.

In most cases, taking the money with you (especially if you plan to buy a car in the automotive market) is not recommended. This is especially for those, who are going to the market for the first time for the purpose of gathering information.

First of all, we shouldn't forget that any market including the automobile one, is a WORKPLACE for professional thieves and fraudsters. And the best protection against them is empty pockets. At present, they have a pocketful of wide variety of techniques and tricks, with which they lure out unlucky car buyers to steal their money.

In addition to that you can easily lose money through pickpockets. For example, when you are taking out a pack of cigarettes from your pocket or a pen and a notepad (to record the seller's phone number) or any other item for that matter.

Also, the lack of money with you can play a decisive and positive role when it's advisable to abandon the purchase of an inspected car. It is no secret that many car sellers are inherently good psychologists and know how to quickly persuade a person to buy a car. At this point, the buyer may not fully realize that it is better to refuse the car as the seller chirps something over his ear and if the car looks decent on the outside. Since he has the money is in his pocket, he makes a decision to purchase right away. And in case if the buyer didn't have the money with him, it would probably have ended like this: if he was very much interested in the car he would have taken the phone number of seller, and back at home again he would have considered the expediency of the purchase.

However, it should be noted that that in some cases having the money with you can play a positive role. For example, many sellers agree to reduce the price when the buyer tells that he has the money with him and he is ready to pay at the spot. Moreover, the size of this discount could be quite noticeable and the seller would never offer it to the buyer who doesn't pay him readily.

In any case if you are going to buy a car and taking the cash with you, take someone with you whom you trust and split the sum between him. It doesn't matter if he has no idea about cars, the important thing is that you should not walk alone with a large amount of money in your pocket.

HOW TO DEAL AT THE MARKET:

Keep in mind that the first trip to the automotive market (Fig. 1.3) is for informational purposes only. It is recommended to visit the market the first time simply in order to find out and gather details about what's going on, long before you plan to buy a car (for example, a month and a half before the purchase). This will allow you to realistically assess your capabilities, figure out and adjust your requirements and maybe also think about your need for buying a car right now (maybe it could make sense to postpone the purchase for half a year or a year).

Many future car owners for whom money is not a huge deal or readily available, end up having sleepless nights bugged by the thought of not owning a car and want to buy a car at the first visit to the showroom or auto market. It is important to note, there is no justified point in rushing to buy a car because you want it! Unless otherwise under dire circumstances where the immediate need for a car is inevitable. Choosing and buying a car is a serious step and an important life decision, so you should never rush with this matter.

So, now after you made the first trip to the market and got the first impression of what is being done there. Maybe you might have the knowledge about everything going on

there but you have a little idea about the situation and know what you can count on. By the way, it's better to not just do a single trip to the market but probably two or three. And many buyers usually do something like this, slowly walk around the market and after some time they come back to do it again, watch the cars and slowly start a bargain somewhere. In other words, "they are being cooked in the market soup", and savouring the choices to choose their right car.

When you are completely ready (by the way not just mentally but financially too) and decided your requirements, you can start choosing your dream machine.

How not to be trapped when buying a car through an advertisement

If you intend to buy a car, Then the dealership or the car market aren't the only places you for it but also you could use the internet for finding advertisements. It is necessary to take into account some of essentials which we will discuss in this section.

Psychological aspects of buying a car

In order to successfully purchase a car, you like, it's useful to know a few simple psychological techniques which are discussed below.

Throughout the time you are inspecting the car, ask the seller about a variety of things (you could also add a little positive vibe with jovial questions). For example, "What is the mileage of this car in Germany? Are you really going to sell such a good car?" or "How to contact you if I have questions about the car in?" Etc., In this case, your main task is not just to gather as much as information about the car you are interested in but also to form a psychological portrait of its owner. It is desirable to get more or less a realistic idea of his personality, about who he is, what does he do, where he comes from, whether he has a family etc. It is also recommended to find out why he decided to sell the car (the need for money, the desire to buy a newer car, etc.). In general, try to get as much information as possible about the owner of the car.

TIP:

While speaking with the owner of the car, carefully listen to everything he tells you, but at the same time be sceptical of his words and it's better if you don't believe everything he tells you.

But criticising the car during the entire inspection is not a recommended attitude. You will hardly be able to beat down the cost in this way (at least if the seller has some experience), but it's quite possible to cause him a negative impression of you. In the end, he might lose all interest in you. It's is better to set a positive tone as you talk, saying stuff like "well the car is very attractive, and that if wasn't for this one little spot (scratches, broken windows, worn tires, and so on) it would have been just perfect!"

During the conversation with the owner, try to learn more about the history of the car,

in particular things like, whether it has travelled a lot outside the country etc. It is advisable to find out who used the car, for example, a student, pensioner, middle-aged family man, young girl, old lady, former taxi driver, etc. For instance, if the owner of the car was a young man between 20 to 25 years of age, it is unlikely to be in good condition as young people, usually take poor care of their cars and even drive quite roughly (taking overloads, driving through holes and bumps, use poor-quality fuel and engine oil, etc.). But on the contrary, a person of retirement age treats his car with care. They take a trip to the country once a month, monitor its technical condition and take good care of the car (since it is very hard for a pensioner to buy a car and also they have good patience about things) and try not to overload it.

Also find out if the car was being driven in winter (if not, this is an additional advantage), as well as the conditions in which the car was stored (in the yard, in a guarded parking lot, in a heated or unheated garage, etc.).

Then compare the responses of the owner with your impression of the car from what you saw. If a car that a pensioner used for trips around the country on vacations from time to time which wasn't used in winters and was kept in a heated garage, does not start well and has rust stains on the body, you are clearly trying to be fooled. Most likely, a young guy probably a student, used it every day both in winter and summer and had no idea about how to properly drive or store the car. He was probably also not familiar with a concept called "car maintenance".

Try to make the seller feel sympathetic towards you. Of course, he will in any case emphasize his good attitude towards you, but it is better that this attitude be sincere and not contrived. Try talking to the seller, but again remember all his words are critical and focus not on what he says, but on how he says.

ATTENTION

Do not forget the fact that everything the seller says and does has only one goal, to sell you the car as quickly and as expensive as possible by carefully hiding all its flaws from you by carefully emphasizing its advantages (even if they are few and sometimes doubtful).

LET US GIVE TYPICAL EXAMPLES

Example 1. The car being sold has air conditioning which is not functioning currently. A foam test will figure out if there is any leakage, and if everything is okay you can just attach everything and it would work. Such a problem can only be deciphered this way, but if the air conditioner is faulty it requires serious and expensive repair. Foam may be bubbling/flowing out because it was rotten or rusted in one or several places and it is possible that this happened years ago.

Example 2. Here is another typical situation: The seller confidently assures the buyer that the car is in excellent condition without any hesitation. But there is a little knock with a valves. „ It just needs to be adjusted just a little, but there was not enough time" he adds with an innocent tone. The only honest and logical explanation for his words are: a valve that knocks is bent and requires replacement, to do this it is necessary to remove the cylinder head which will require additional (and considerable) money.

Example 3. Another famous example. During the test drive, the buyer notices that the vehicle „ pulls „ to one side. The seller argues that it is necessary to adjust the camber and toe-in which is a cheap job at any service station. In reality, there might be a possibility that this car has previously happened to be in a traffic accident, as a result of which its geometry was disoriented, or at least the suspension was badly damaged. Therefore, in such cases it's better to put a condition to the seller that "we are now going to the service station, adjust the camber and toe of the wheels there after which we make a test drive and if the car drives normally then I will agree to buy it". If the seller does not agree it is better not to waste your time with the car.

Thus, we formulate a conclusion from the above scenarios that, when inspecting a used car do not believe what the seller says but your own eyes, sensations and experience. At the same time, it's not advisable sharing with the seller all the doubts and conclusions you have because an experienced person will easily find a truckload of various excuses and explanations desig ned to confuse you and believe in the veracity of his words.

If you are buying a car through advertisement or from the automotive market be sure to bargain. Is there any market where buyers do not bargain with sellers? To be honest, it makes perfect sense to bargain even in the car showroom too, although many mistakenly believe that it is useless. I emphasize once again that you do not criticize the car as a whole, but clearly and specifically point out to the seller the disadvantages it has. Tell him things like, you have dreamt about owning such a machine all your life and everything suits you, but it's bad luck that you do not have enough money and it's impossible for you to get the rest of the amount ready literally by tomorrow or the day after. If the seller yielded a little the car even be brought right away for a cheaper price. By experience, this simple technique in many cases can significantly reduce the price of the car.

When buying a car on the car market or through an advert, remember the important rule: If you don't like the seller for some reason or it makes you suspicious about him, and if any suspicious persons are turn around the car being sold, immediately refuse to buy, even if the car itself has no complaints and completely suits you. Otherwise, you can be caught in a web of serious troubles and even become a victim of a crime and robbery, not just fraud.

You have every right and reason to doubt the honesty and decency of the seller if he clearly holds back something, hides something from you, if he tells a lie even if it doesn't directly concern the car, if he leaves the conversation and suddenly starts to rush off somewhere for no apparent reason.

As we noted previously, it is necessary to realize that the seller's smile and friendliness are solely from his desire to sell his car as soon as possible at a better price, not because he is glad to see you. He is more concerned about his task and you your own, so in this situation you both are clearly on opposite sides and in other words you both are rivals. The one amongst you who is wittier, persevering, persuasive and dominating will be rewarded. But, do not rely on trying to trick him, but firmly and adamantly hold on to your words and mentality that "I want to buy a normal car at a reasonable price and I can assess the advantages and disadvantages of this car without the help of the seller".

In no case should you inform the seller in advance about which particular car you want to buy and even more the details of how much money you are ready to pay. Without a doubt an experienced seller will be able to convince you that his car is what you need. Even if you wanted to buy a car which is no more than 5 years old with a "Universal" body, for less than 10,000 euros, but you stupidly told this to an experienced seller offering a 7-year-old car with a 'sedan' body', after being influenced by his words, you will quickly reconsider your views and buy the particular car. And when you realize that you hurried up and bought something which you didn't originally plan for it would be too late.

Before going to buy a car, carefully read the newspapers and magazines devoted to the sale and purchase of cars in order to have a more realistic idea of the price range currently prevailing on cars of the brand and model you are interested in. It is also recommended to talk about the topic with knowledgeable people. Pay special attention to the existing shortcomings of the model and make of the car you are interested in; this knowledge will be useful in the process of bargaining with the seller.

After inspecting a car that interests you, if you came to the conclusion that it suits you perfectly, rather than asking the seller how much it costs it's better to just tell him for how much you would agree to purchase it. And when talking with the seller, clearly but subtly let him know that you cannot spend more than the amount you mentioned.

It is best to buy a car during the end of the day, during the end of the working week, during the end of the year or on the last day just before a long holiday season. For example, one of the most attractive options is the last working day of the outgoing year. If this is not possible, you can simply do it during the end of the working day. At these times the likelihood of a significant price reduction is much higher because at the end of the day the seller is tired and at the end of the year the idea of selling a car and quietly going on holidays (for the week-end) looks very tempting to the seller which is a huge plus for us.

In the process of inspecting a used car if you find it has several flaws, you shouldn't tell the seller about all of them at once. Instead each defect should be discussed with the seller at different consecutive appointments which increases the chances of the price being discounted in every visit.

THE MOST CHARACTERIS-
TIC MISTAKES OF BUYERS

Despite the fact that car buyers represent different classes of our society, most of them make the same mistakes when choosing and buying a car. We will analyse the most typical such errors in this section.

What kind of tricks and techniques preowned car dealers use to give the car a glossy, shiny and fresh look? Here they usually do repainting, polishing, and general cleaning of the cabin by special means and polishing the plastic elements (dashboard, etc.). A special oil "specifically sold for used cars" is poured into the engine, which hides the noises in the engine. Also new wheel caps are fitted making them look new, tuning is done, etc.

The most common mistake of all buyers is that, they make hasty conclusions and are subconsciously are ready to buy a car that externally makes a good impression (Fig. 1.6). The very important quote which you should recall here is "All glitters are not gold" and therefore, cars prepared for sale should be treated with a high degree of criticism and caution.

However, there is another side to the coin. If the car looks great, it's highly likely that the former owner treated it well and constantly monitored the condition of the car.

A characteristic psychological feature of almost all buyers is, after arriving at the car market or the store, they have a positive attitude towards almost all cars put up for sale. It seems to them that all these cars are worthy of buying and the only problem is to choose what they need from this abundance of choices. At this instance, the person completely forgets the fact that car being sold may be listed as a hijacked one, has problems with customs, or the numbers have been changed and also that it could be a rotten old dead car which is presented shiny perfect on the outside.

And here is another typical mistake that is very common in beginners. The idea here is that, when choosing a car, a person is guided by a number of completely different requirements which should be taken into account.

This is especially true in the case of women. You can often see a conversation with them like „ What kind of car do you want to buy?" which is followed by an answer like „ White!" (Options - red, green, etc. colour based on the lifestyle). Remember, you need to choose a car primarily on the basis of what you need it for (for driving to the office, for trips to the country house, etc.) and as well as on the basis of its current technical condition. Otherwise, it may happen that the white car which you bought with modern features and all the „ accessories" would breakdown before you reach home in it.

It is not very uncommon for a used car to be equipped with a buffed up stereo system which can cost almost the price of the car itself and it could often be a decisive factor for the buyer who is a music lover. It is recommended to realistically assess the situation and decide for what you intend to pay the money for, either for the "stereo system" or for the car. Even if you are a big connoisseur of music please do not forget that you will always have time to buy any good audio system separately in the car market or in the store.

IT'S IMPORTANT TO NOTE

The price of the audio system in the car falls as quickly as the car itself! (it's no secret that every car gets cheaper every year).

Many inexperienced buyers are trying to save money by buying a car with modern audio system (which is usually a trick to make them buy) and think that this equipment can be sold and something cheaper could be purchased instead, thus they could get could get back some of the money they spent on the car. This is a common misconception and even if this plan was really profitable, the seller would have thought of it a long time by now.

Don't forget that even the choice of the model and brand of the car should not be approached dogmatically. For example, you intend to buy a car with a body type „ universal". After analysing things such as the availability of your country house and assessing the amount of money you have in your hands, you decide that such a purchase would be optimal. However, you couldn't find a suitable car of such a brand on the car market, either the body is not in the best condition, the engine knocks or it is expensive, etc. In such a situation it is advisable to look at a car of other brands, but of a similar class and year of manufacture. For example, "Ford Escort "of the same year of release and also you could see other cars of the body type „station wagon". These cars are about the same class and are in the same price range. In terms of performance, such cars are also same in general.

For many drivers a key criterion is the presence of an air conditioner. Of course, it is very convenient when driving in hot weather with an air conditioner and the air conditioner also does an excellent job defogging glasses. But despite all advantages, the availability of an air conditioner should not be the decisive factor while choosing a car.

The only exception is when the car is supposed to transport people suffering from cardiovascular diseases (for them, driving in the heat could be torturous or even fatal).

Also, many future car owners attach too much importance to having a full power package such as power windows and central locking system and they simply don't consider the cars that don't have them. Needless to say, this is a completely unreasonable approach! In the end, if you are not disabled and your arms and legs are perfect, you can open the window, lock the door or adjust the mirror manually.

REMEMBER THIS

Air conditioning, power accessories, audio equipment and other fancy equipment are only extras to the car and nothing more.

Another very typical mistake made by car buyers is not buying a car according to the needs, but for a show off like "I want a car like a neighbour (friend, brother, matchmaker, etc.)". To be honest, it even surprises how adults can make such a responsible decision (which is undoubtedly the decis-

ion of choosing the car) on the basis of a little kindergartener attitude like "I want the same on as the other kid".

This may lead to a situation that the purchased car does not satisfy your requirements in any way. For example, a young unmarried man who doesn't have a family buys a minivan for himself, which meets the requirements as the same as that of a large family. After some time, the hapless owner of the machine is aware that he bought something which is not what he wants, as the vehicle consumes a lot of fuel, there is too much space on the inside where there isn't anybody to use it, a huge trunk which is empty all the time and it's hard to manoeuvre such a huge car in the city. After some time, this guy comes to a logical conclusion that he doesn't need a minivan but a small three-door sedan or hatchback model that was specifically designed for his needs.

There are also situations that are entirely opposite. For example, a middle-aged man with a large family and who owns a summer cottage buys some three-door small car which is probably the same as that of what his neighbour drives to the supermarket. The only reason he has for choosing the car is "this machine is so pretty and manoeuvrable! „. And then he starts to face the problems as the large family could hardly fit in a small car, the trips to and from the cottage turn into pure torture because it is impossible to accommodate everything you need into a small car (especially in the autumn when people load their cars with bags of potatoes and other crops).

If you are a happy owner of a summer cottage and have a big happy family, then you need a car which isn't just roomy, but also with a high ground clearance. This is primarily due to the fact that Russian suburban (and indeed many urban) roads are notable for being bumpy, has a large number of potholes and other similar defects. For access to many Dacha cooperative zones unpaved roads are used and mostly in these areas there isn't any roads probably, but just road paths which are really bumpy. Therefore, on vehicles with a small ground clearance which are being used for frequent country trips, the underbody often is damaged due to all the bumps. And do not forget that the trunk of the machine is always loaded during travels which further lowers the car down and decreases the ground clearance.

By the way, for most farmers the best car are the ones with body type „universal" and with five doors. Such cars have a trunk which extends behind the rear seats as a continuation of the passenger section and the fifth door is located at the back, designed for loading / unloading the goods transported.

The rear seats in these machines can be folded so that it becomes a continuation of the trunk making more room for even some large loads (for example, a refrigerator) to be transported. Also, many summer vacationers prefer minivans, but however cars with this type of body are intended primarily for those who like to travel by car.

Many buyers (especially for inexperienced drivers) who intend to buy a car, completely forget the fact that every car needs maintenance or repair from time to time. And the cost of spare parts and repair work for cars of various brands can be very different. Often this fact directly affects the price of a car because a car whose maintenance and repairs are expensive could be cheaper than a car of the same class which requires less money to maintain.

TIP

If you have purchased a "second-hand" car, you should not go on a long trip immediately after the purchase. At this early moment you shouldn't have too much confidence in its reliability. You should check the car by driving within your settlement for some time and do the necessary maintenance (oil change, filters, general inspection of the condition of the car, etc.) before you plan to take it out on a long journey.

It is recommended to take into account the upcoming costs of the maintenance and repair before purchasing a used car. Otherwise you will land in a situation where you cannot use the purchased car as you will need to repair it after and you might be short on money. Experienced motorists recommend buying a used car for such a price that once you have purchased the car you have some money left in your pocket which is roughly about 10 - 15 % of the value of the car. In other words, if you purchased a car for 10,000 euros, then you should at least have 1,000-1,500 euros left in your reserve.

Another characteristic mistake of many Russians (usually this applies to wealthy people who aren't used to counting their cash before they spend) is an irrepressible desire to purchase a car of an exclusive brand which is almost never encountered and very exotic. Probably through this way they prefer to flaunt their originality. However, almost the day after the purchase, it becomes clear that such originality is associated with a huge number of unnecessary and completely unnecessary worries and problems.

As we noted earlier, every car breaks down once in a while and it has to be serviced and repaired. This is to be noted as finding parts, components and consumables for a rare car is very! Very! difficult. On a rare car the probability of finding these things are very less as they will not be available in the store, so you will have to order them from abroad (this service is offered by many shops and car dealerships). In the end, the service which is commercially available and relatively cheap for an average car of the same class will prove to be a truly expensive for you.

In addition to problems with spare parts, you will have to think about where to repair and maintain this rare „ exotic machine „. If there is a company service centre in your city that deals with such cars, then everything is more or less normal. Because trained professionals will deal with your exclusive car. Of course, it will cost a lot of money but as they say - "You pay for what you get!". Things can get really complicated if there is no such centre in your locality and your surroundings, and your rare car will have to be given to the local mechanic. In general, such experts do a good job with ordinary cars but they can seriously spoil a rare and exclusive car due to the lack of specific skills.

And further, it's known that many foreign cars are very specific in terms of fuel quality and do not like Russian fuel so much because of its low quality. So applying this fact to the rare and exclusive machines it can make the situation even worse.

One of the most well-known mistakes is that, a person expects a good quality car for a low price. The desire is quite understandable, but you should realize that "you get free cheese only in a mousetrap", so buying a reliable and durable car for a cheap price is almost impossible. In other words, if you intend to purchase a car, align your financial capabilities with your requirements.

A common situation is when a person wants to buy a car of a famous and prestigious brand for the modest price. However, he has an idea which is something like" It doesn't concern me if the car may not be in the best technical condition! The important thing is that it's a 'Porsche' (Lexus, Mercedes and so on.)". It's a very absurd unreasonable approach and if you are guided by such considerations, you could end up buying a broken and totally destroyed car which might only have its prestigious brand name left.

It's no secret that the cost of certain models and brands never go down and almost always lie in a certain price range. The same old Mercedes W124 / 280E can cost as much as 6 or 12 thousand euros today. This stagnation of price is the cause of another known error. Usually a person wants to buy a car at a price as close as possible to the lower price limit. Such savings can be understood from the practical point of view, but you should know that the cheapest cars are either in poor useless technical condition or have a bad history (stolen, not cleared by customs, etc.). But if you are going for a car at a decent price which is least above the average price for cars of this brand and model, you have every reason to expect that the car will be in very good condition and will not require any serious expenses in the near future.

HOW TO PICK UP A CAR?

MY FREQUENT ADVICE

AT FIRST

1.Accurately determine your budget. You are wasting time and energy looking at a car worth 10000 euros if your limit is 5000. You should also know exactly how much money you plan to spend. If the car is in good condition the seller will not trade it with you for half the cost.

2.Have 2-3 sources from which you can search and keep track of advertisements. So you will have more time to look into the advantages and disadvantages.

3.Choose only those cars which you can easily repair and service. It's not worth chasing a 15-year-old BMW, even if you could afford it at the price of a Skoda, because servicing the BMW can cost about 2 to 2.5 times more expensive. It is crucial to assess the problems with the BMW which could possibly burn a hole in your pocket.

4.The clearer idea you have about the specification of the car you want (specific engine, gearbox and powertrain), the easier your search process will be. Yes, customers often select without any specific requirements in mind and in the end they spend a lot of time trying to choose the model, which eventually leads to frustration.

5.Do not wait hoping that someone will sell good cars at a cheaper below average price. If the person urgently requires money, he sells the car through intermediaries and receives the money. All other cars you find at a cheap price either have some issues, broken or have legal problems. This could be a possible advertisement by scammers too.

6.Determine the average market value of the car which you are looking to buy. You can either search it yourself or you can use the built-in functions on popular car advertisement sites. If the price is greatly less than the market value, then it should be having some problems which can incur additional costs or it could be a stolen vehicle with missing documents. In any way it should be doubted.

I'LL SHARE A CASE WHICH I ENCOUNTERED

A person approached us with the following situation: he received a letter from the court that his car was subject to forcible seizure. The person clearly began to panic as he had no debts and had no idea why his car needs to be seized.

LET'S START FROM THE BEGINNING

About a year ago this person, let's call him Valery for namesake bought a car from a private trader and the seller provided him with all the documents. So everything should have been ok, what's the problem? It turned out that the owner of the car was already subjected to the seizure of the car at the moment of purchase and this was found out only later when we started to dig into the details.

Many of you might be confused how this situation can even happen! will explain everything in detail on this example:

We will call the seller Gonza, who took a loan from a bank long ago and as the moment of repayment came he simply understood the fact that he could no longer pay the bills. So he reached the point when he exceeded the time limit and was registered in the database. And then he starts creating problems for others. Gonza starts selling all his property, but instead of covering up his debts, he stupidly dumps them into another city or country. But he has already been entered into the database, which means all his property already does not belong to him. But of course our guy Valery had no idea about it and when he found a great car for a very attractive price, he immediately purchased it.

WHAT COULD BE DONE IN THIS CASE?

I will literally tell you what the Jury told us "Stand in line guys, you are the fourth in the list".

Go to law? To whom? to Gonza? who knows where he is? Get into suing one of the largest banks in the country?

CONCLUSION

If you are going to buy a car it is not important from whom you buy it. It could be private owners, car dealerships or dealers, don't forget check its history and don't hesitate to spend a couple of bucks to check the owner on the seizure register. Unless of course, you want to play the lottery with the just purchased machine.

7.Examine and research about the model you are looking for. Often the variant you like can be quite problematic.

8.The best cars at reasonable prices are bought literally within a few hours or at the maximum in a few days, so if you find a good one do not slow down or postpone the meeting until the next day or until the next week. Here's an example:

A girl came to me with a request of finding a car. Her criteria were very clear. We began to search for the option that fits in all her requirements. We reconsidered many options, but as it always happens: the car was either not in the best technical condition or the other features of the car did not suit the taste of the client and in general we were looking for quite a long time.

SO

Literally a week before, I got an offer to see a car from an acquaintance of one of my acquaintances, on the recommendation of some acquaintance (some kind of tautology turned out). The owner of this car turned out to be a grandfather who actually wanted to sell it at the car dealership but he was offered quite less and they especially started to find different non-existent breakdowns (as we knew later).

We agreed to meet, so I met with his grandson and I just went nuts as the car was from 2011, bought in the Czech Republic, 1 owner, hasn't done even 10,000 kilometres completely (which was the true mileage! I ran a lot of checks!) and was in perfect technical condition. In fact, the car is „brand new" in factory paint except for a few little scratches which are just minor external defects. Well, everything was clear and this is just the option that the client needed (it almost suits all the requirements). I sent her a photo and a reply came saying that "I want this colour".

Everything was fine, but they only agreed to show me the car and especially since the grandfather treats foreigners with caution I met with the grandson.

COMING TO THE POINT

I knew it perfectly if I just leave now, I could simply miss something really worthwhile. After all, this kind of option is one in a million opportunity and the price was attracti-

ve. The price doesn't matter much because such cars are hard to find for sale. I took a rather risky decision, I suggested to the grandson that he contact his grandfather and inform him that I give him a pledge and come back tomorrow with money to buy the car from him (I strongly advise against doing this normally, but this was a slightly different situation).

So on the same day I called the client and explained the whole situation. I told her that she should see the car tomorrow and finish the deal or I'm ready to buy such a good car myself.

The result was: The client wanted to buy the car and it was a happy ending:)

9. Never transfer money to anyone as a deposit without a receipt.

10. Remember that car dealership and even official dealers sell broken cars casually without any hint. I have seen several cases where they claim the car to be unpainted but the inspection would prove otherwise.

11. "Grey(fraud)" car dealers often carry out frauds with contracts, so carefully read each page when you bring a document for signing.

12. A car with a clean engine and engine compartment should be checked with a higher degree of caution. Probably the engine was washed either out of ignorance or to hide any traces of leaks.

ENGINE CHECK

When inspecting the engine note the following points: The engine must be dry and clean but not freshly washed! Washing with a spray gun is possible to give a good clean look. Therefore, look more closely at the motor for oil leaks. Inspect the hoses, they should not be cracked. Assess the condition of the oil, presence of small metal particles indicates the wear in crankshaft liners and water bubbles (antifreeze) indicate a damaged head gasket. If the oil is fresh and very thick, then it may have been replaced recently to hide any serious malfunctions (low compression and various types of wear). Open the oil filler cap and inspect it, scratch the inside of the cap with a screwdriver. The presence of any rust indicates that the engine had overheated from time to time. There should be no black pasty coating inside the neck. The presence of black sludge is a sign of poor engine lubrication, either the oil of the wrong specification was used or it wasn't change for a long time or the engine overheated. Inspect the inside of the filler neck and the valve mechanism with a flashlight. In a good engine, everything you see inside should be golden brown in colour. Carefully inspect the cooling system for any leaks. Check whether the hoses retain their elasticity. Open the coolant expansion tank cap. The presence of rust in the tank shows in the fact the engine has been overheated previously and the presence of oil in antifreeze means that the "facing" of the engine head or block has been damaged or the head gasket has been damaged. Few rules for inspecting a running engine: Start the engine. Before it warms up, open the cap of the expansion tank. If bubbles rise, the head gasket is damaged. Carefully open the oil filler neck and if you see a lot of gases go out then there are piston ring defects or crankcase ventilation is damaged. Inspect the engine for leaks or leaking of antifreeze. Listen to the engine, there should be no „metal clanking" sounds. Engine speed should not keep "floating" and be steady. On a warm engine, ask the assistant to press the gas pedal abruptly and firmly, and at the same time place your hand over the exhaust pipe. The presence of a large number of oily spots on your hand indicates, minimum to significant wear of the piston rings.

CAR INSPECTION

13.Always try asking as many questions as possible through the phone.

14.Ask just not general information, but specific ones. For example, when was the last inspection, where it was carried out, what was changed and so on.

15.Memorize or even record the answers. You can verify if they tell the same when you are inspecting the car in person.

16.Only the owner should sell the car in person. No godfather, no brother, no matchmaker. In extreme cases, the wife's car can be sold by the husband. In these cases, check the stamp on the vehicle passport so it's valid.

17. Check the VIN number on the car. This is very important and many skip this step assuming that they would not be cheated.

18.Ask the seller the things you asked by phone and compare the answers.

19. If it's possible, rent a thickness gauge.

WHAT IS A THICKNESS GAUGE?

Thickness gauge is a device used to measure the thickness of the paint coating. It looks like a little box. Some thickness gauges work at temperatures below zero and others do not. Some models measure aluminium parts, others work only on steel surfaces.

I recommend using professional thickness gauges.

What does the thickness gauge show and how to use?

Using a thickness gauge is simple. Take a calibrated device(gauge) and attach it to the car body at a right angle. It will immediately show the coating thickness in micrometres(microns). One micrometre is one thousandth of a millimetre.

The greater the value indicated by the thickness gauge is, the greater the layer of paint and there is a high chance that this car was repaired and repainted after an accident. Usually the factory paint layer on steel parts of a car is no more than 200 microns.

Indications above 200 microns indicate signs of repainting.

Indications up to 300 microns occur if any small surface defects were painted over, for example, a scratch from a key. This does not affect the safety of passengers, but you can use it to bargain the price. Values closer to say 1000 microns indicate that there is putty under the paint. Such bodies had been dented strongly or deformed pretty badly in the past in an accident probably. If the body restoration work was done poorly, the paint on the putty can crack and fall off over time.

More than 1000 microns is a sign of serious body damage and therefore indicates the car had been in a serious accident. It is better not to buy a car with a coating layer thickness of more than 1000 microns.

2000 microns is the maximum value that the thickness gauge can show. If the layer is thicker, the device will not show numbers. This means that lots of putty has been used.

Check the values of the factory paintwork thickness of the car which you are going to purchase in advance. This information is easily available on google, for example with a keyword like „BMW 3 series paint coating thickness". Assume a small difference in value is fine, but not more than 60 microns.

HOW TO CALIBRATE A THICKNESS GAUGE?

To show accurate results, the thickness gauge must be calibrated. The values deviate due to temperature fluctuations or if the battery Is low.

To calibrate the thickness gauge, use the metal plates that come with them in the kit. If you rent a gauge, ask the owner to calibrate it with you and get a calibration plate from him, just in case.

Calibration plates look like this:

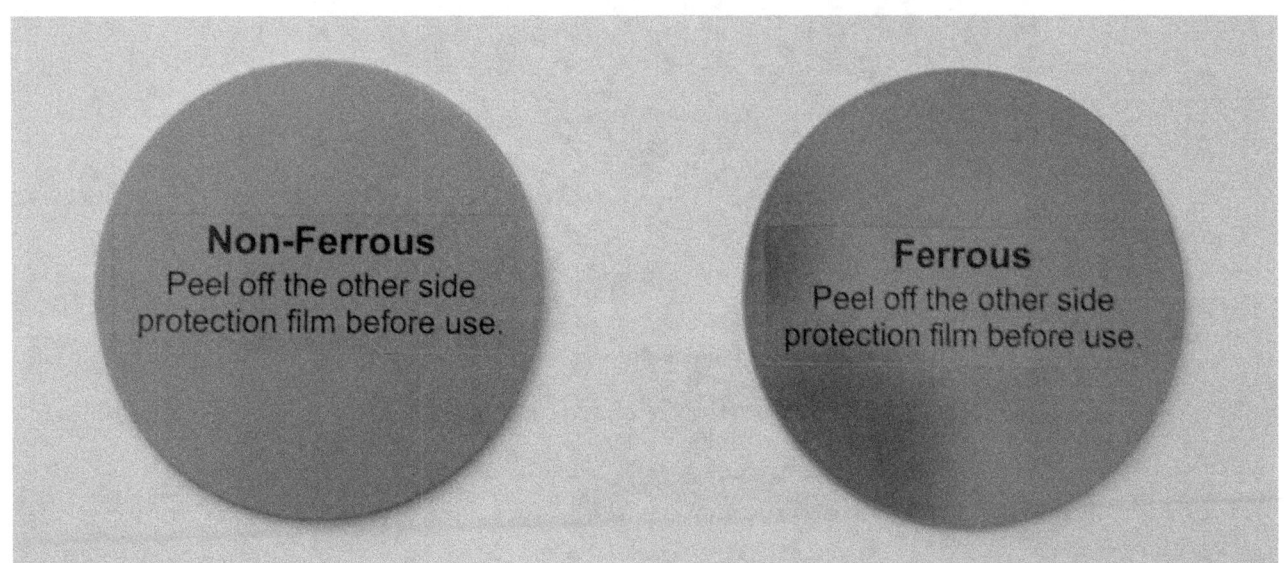

Non-Ferrous
Peel off the other side
protection film before use.

Ferrous
Peel off the other side
protection film before use.

My thickness gauge has two calibration plates, as it works on both steel and aluminium parts. Calibrate the device separately for steel and aluminium. If the thickness gauge works just on one metal, then there will be just one calibration plate.

The kit usually includes a special calibration film, which is the reference.

The calibration process is simple, we place the thickness gauge on the plate and reset the readings:

We attach the calibration film on to the plate and place the device on top:

The thickness gauge should show the numbers printed on the film. Repeat the procedure until you achieve this:

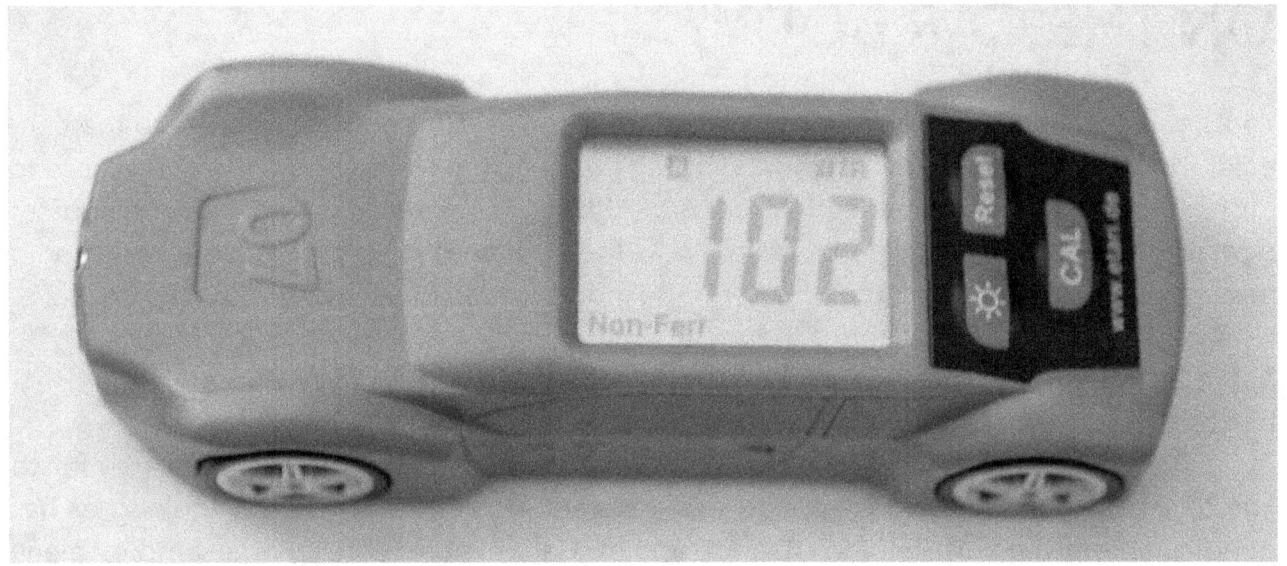

(Car paint coating gauge MD 07 from Etari GmbH, Stuttgart, Germany - www.etari.de)

WHAT ARE THE THINGS TO CHECK ON THE CAR BODY?

You need to check at least five points on each body element. Body elements are for example: a car roof, bonnet, fenders or door. Each element is checked by the thickness gauge at the edges and in the centre. Be sure to check the pillars, this is the metal frame between the doors passing from the bottom panel to the roof.

The thickness gauge must be applied perpendicularly to the surface. If you attach it from a different angle, the readings will be higher or different.

It could be that just a part of the car body has been repaired and painted. Places of relief on the car body design such as places where there are curves and sharp ends. Check everywhere around these areas with the thickness gauge including above and below the curves otherwise you might miss some areas. For example, you may check the top of the body panel but the putty must have been somewhere below which you missed.

Measuring the outside of the body for thickness is just half the story. To avoid any repairs in the future be sure to check the inside of the body by measuring the paintwork on the frame as this would indicate if there has been any damage incurred to the frame in the past.

Open the door and gain access to the centre pillar. Open the hood and get access to the fender shelf and cups. The readings of the thickness gauge in these areas should not exceed 100 microns.

ALUMINIUM PARTS

Before inspecting the car, it is helpful to know if the model you are looking at has any aluminium parts. For example, many models of brands such as Audi, BMW, Range Rover, etc. has parts made out of aluminium such as the hood, front wing(fenders), doors, etc. To measure the coating on these parts, you need a thickness gauge capable of working on aluminium. If the car has an aluminium part, but the thickness gauge recognizes it as steel, then the part has been replaced and not the original factory made one. This is another sign of accident history of the vehicle.

PLASTIC PARTS

The thickness gauge does not work on plastic parts. Modern cars have bumpers, door handles, exterior mirror housings and sometimes other parts made out of plastics. For example, the Peugeot 408 has plastic front fenders.

It is better to study the features of a particular car model before you inspect it. This can be done by watching a few videos about the car model you looking for on YouTube.

Using a Thickness gauge on a dirty car can show false reading and it could be damaged.

Using a thickness gauge on a dirty car is a very wrong step. Firstly, it will show incorrect readings. Secondly, using the thickness gauge on a dirty body causes damage to both the device and the paint of the car.

Wash the car before doing checks. Make sure you inspect a clean car. This helps you to inspect with higher accuracy. You can find more information about the thickness gauge on the internet.

Remember that the Thickness gauge reading doesn't completely guarantee the quality.

This is because after a moderate accident, parts from other vehicles can be replaced on this vehicle. During the repair if the mechanic could find parts of the same colour from other cars which can be replaced to look like original, the thickness gauge will show original value on the external body elements.

Therefore, thickness gauge inspection isn't just enough. While inspecting you need to pay attention to the gaps between the body panels, the condition of the bolts, the year of production of glasses and a lot of other parameters.

20. If it's possible you can rent a device to diagnose electronics. If you cannot, do not worry you would still can check it as a service station.

21. Compare colours and shades of paint on the adjacent body panels.

22. Check for any smudges of paint.

23. Check for dust and hair in the varnish.

24. Look closely for any shagreen patterns (uneven coverage, like ripples on the water) on the adjacent parts. Their reflection should be the same.

25. See if there are any "Cracks" on the body and specs of rust. „Cracks" are barely noticeable cracks in the paint.

26. Inspect the mid parts of body panels. Often the tone transitions during painting are made there as they are unnoticeable.

27. Take of the rubber seals and covers to see if there is any difference in the colour of the paint under them.

28. Look at the gaps between panels, they should be the same on the left and right sides of the vehicles.

29. Inspect the car in good daylight or preferably in cloudy weather from different angles. Any small dent or scratch is a reason for bargain.

30. Inspect the paintwork of the machine very carefully at each and every place for microcraters. Microcraters are grooves no larger than a millimetre in size in the paint or in the varnish.

31. Clearances should be uniform along the entire length.

32. Check the labelling and manufactured date of each glass (windshield and window planes). Vehicles manufactured in Asia will not have the date of manufacture, but you can check the numbers on the glasses.

33. Check if the stamping and mouldings fit perfectly on different parts. They are very difficult to be fitted perfectly once they have been removed.

34. Look at the bumper. He should not speak for the body.

35. Look at the headlights and foglights. They should have the same labels and they should be equally cloudy.

36. Look under the floor mat inside the trunk. There might be traces of paint or damage as usually nobody masks those places.

37. There should be no water in the spare wheel space.

38. Look at welds and spot welding zones. They must be symmetrical and identical on both sides.

39. Check for tire wear. They should be evenly worn out.

40. Check the main frame and sub frame frames as these are the main strength structures. There should not be folds, traces of paint or any other defects. (If you do not know what frames are, look at the pictures on the Internet, this would give you a better explanation.)

41. The engine compartment should not be washed and there shouldn't be any traces of leaks. It should look normal.

42. See if the paint is etched out near the bolts on the doors, hood, tailgate.

43. Check if all the bolts used are the same and standard ones.

44. Look for any faded paint on the sill and around the doors. If there is any, check if there are similar marks on the door too. If it's the case, then the door rubs against the frame and this is not good. If this isn't the case then these are probably scratches from the heels and bags, everything is fine.

45. Pay attention to the fuel tank cover flap. Paint on the bolts shouldn't be etched. This is because most people use the fuel cap colour as the reference to select the colour for repainting. If you see scratches it means the cover has been removed for this purpose.

46. Check the shock absorbers by pushing down the car from each corner so that it oscillates. There must be only one up and down oscillation, otherwise shock absorbers need to be changed which is quite expensive.

INSIDE THE CAR

47. Look at the wear on the steering wheel, pedals, gearshift lever and armrest. This can give you an idea of the actual mileage of the vehicle. Though the wear can vary from vehicle to vehicle you will be able to figure out soon with some experience.

48. If there is a steering wheel cover, remove it to assess the real condition.

49. Try sniffing the cabin out. It shouldn't have a damp smell to it. There should not be any foreign odours.

50. Be alert if it smells strongly of perfume or room freshener. Most likely, the seller wanted to mask some other foul smell to make the car presentable.

51. Check the upholstery under the seat. It should be dry and should not crumble out like a powder.

52. Check under the floor mats. The carpet under the seats should not be damaged.

53. See if there is a dent under the mat due to the right foot. If there is one, then the mileage is far beyond 200 000 km.

54. Look at the upholstery. It should not look swollen and weathered out.

58. There shouldn't be any cracks on the dashboard.

55. Look at the seat belts. They must have plastic buckles around the metal plate that is inserted into the lock. If they don't, then most likely the car was in an accident.

56. Check the working of the air conditioner. It is best to check the paired metal tubes that go from the engine compartment to the passenger compartment (one must be cold and the other hot), but you can simply turn on the air conditioning at maximum fan speed and minimum temperature to do a simple check.

57. Check whether all the warning lights on the instrument panel are working when the ignition is turned on. Usually the all the lights glow once when you switch the ignition on.

58. Oil pressure indicator and airbag indicator should go out separately.

59. After starting the engine, all the indication lights should go out (except for the parking brake, if It has been engaged).

TEST DRIVE

60. Check the oil level. The level must be between the "minimum" and "maximum" marks. Oil should not be black (especially if it's a gasoline engine).

61. Oil should not smell like fumes, there should not be any particles, sludge or debris.

62. After starting, there shouldn't be any thick smoke from the exhaust pipe.

63. If the car is a manual, check the clutch. If it engages right away, its fine. If not, the clutch will have to be changed soon, and these are additional expenses and a reason to bargain.

64. There shouldn't be any jerks when switching gears from P to R or from R to D and back.

65.There should be no jerks when switching gears while driving with the accelerator pedal completely floored.

66. The car should not pull towards one direction while driving.

67. The steering wheel should be straight when the car goes in a straight line.

68. Turn off the music and listen to all the sounds. Memorise any weird sound in your mind and then describe it to a professional for expert opinion.

69. When you slowly brake the wheel should not jerk or sometimes perhaps it could be the ABS and you need to check it.

BEFORE THE PURCHASE

70. Inform your expert assistant about all the weird different sounds you heard during the test drive and ask him to make a service check before purchasing. You should also check the suspension and find out how serious it is, because this will give you reason to bargain with the seller.

71. If you did not/could not do the electronics diagnostics, check it out at the nearest service station.

72. Make a reasonable bargain.

73. Check the car site

74. Check the if the car is recorded under any legal issues on the website of the bailiff service.

75. Check the car on the site.

76. Clearly record the cost of the car and the amount transferred to the seller. Make an official receipt. If any problem arises, then you can go to the court and get your money back legally.

77. Also don't forget to mention that, at the time of drawing up the contract, the car was not sold to someone else, is not under the ownership of someone else, is not mortga

ged, not seized and is not a subject of dispute.

78. After you pay the seller, if he suddenly asks you to terminate the contract and promise to return the money, always check the money in the bank immediately. In most cases you may receive a bouncing cheque or fake promises.

79. Believe only after you have checked it personally and do not trust the sellers.

80. Do not buy a car from a friend, relative or any close acquaintances without checking. They simply might not be aware of all the problems which can lead to quarrelling or bitter experiences between you.

81. If you have a friend who knows cars, do not blindly trust him. Discuss with him these rules and check them side by side.

82. Just choose to do a complete inspection on the car which you like and finalised, which matches most of your requirements. It makes no sense to spend money on the diagnosis of each and every option you have. Plan such that you will end up buying the car which you diagnose given that it's in a good condition.

HOW TO INDEPENDENTLY DETECT HIDDEN PROBLEMS IN THE BODY?

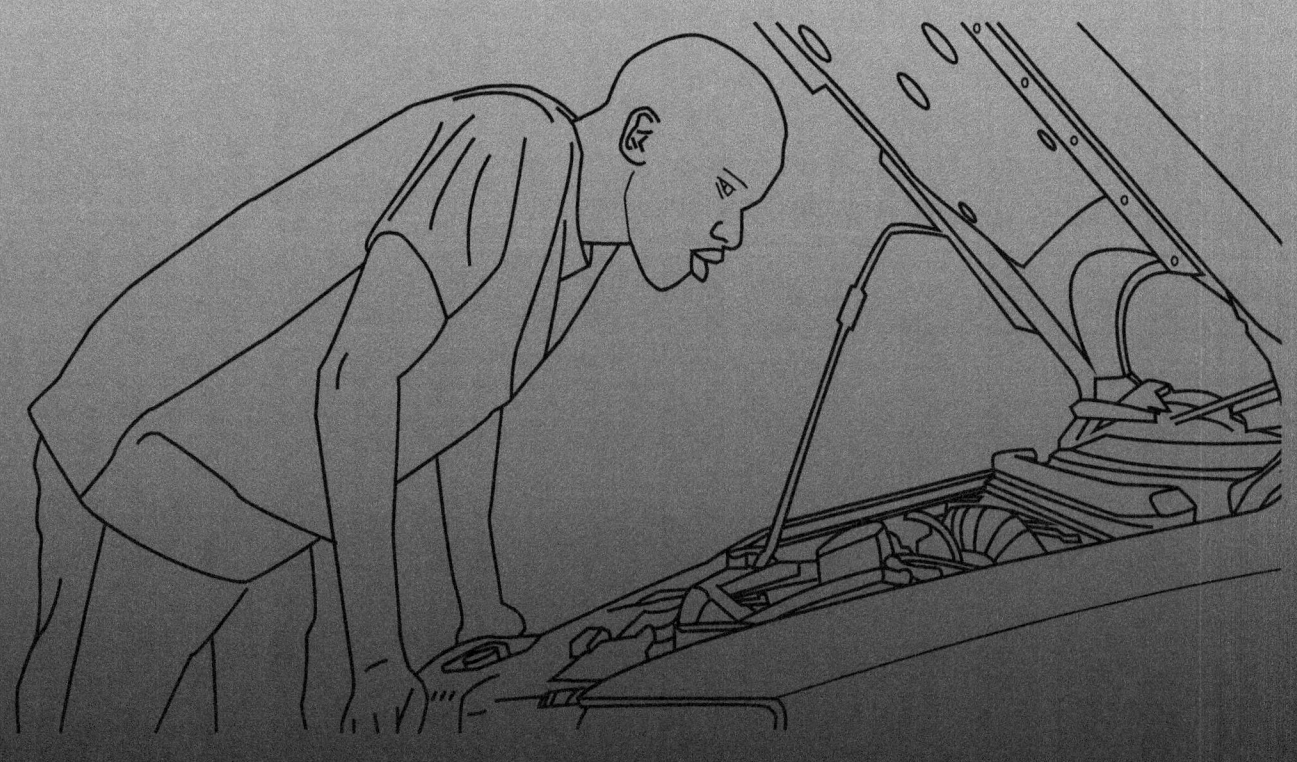

Inspection of any used car begins by examining the condition of its body. And there is a high probability that a car could be rejected just based on its external appearance.

Most people who are new to this have a common misconception that the state of the car's body is not the most important thing. They might argue that the engine, suspension and the drivetrain are the important aspects, but the dents and rust on the body is not going to affect the speed and performance of the car. Such deeply mistaken idea and carelessness can lead you into a lot of headaches after the purchase.

This is because the condition of the car's body reflects the overall condition of the vehicle. It is very unlikely that a car with a rotten, rusty and dented body would have a reliable motor or drivetrain (gearbox, suspension etc.). It is logical to understand the fact that all components wear out the same over time.

An important note: During the inspection, try to make sure the vehicle is clean and well washed (Fig. 2.1).

Dust and dirt on the body can act as a good mask and hide the defects well. If there is a car wash nearby make sure you pay a visit and wash the car completely, even if it means paying additional money to the seller for the washing. Besides the fact that washing can reveal the car body better, it can also help you diagnose the car for any possible leakages in places such as doors, glass seals, under the hood, into the trunk etc.

If the car body is not glossy and has a matte shade, then this is an indication that the car is pretty aged up and is far from being new. It's even worse if the body panels or parts of the car are in a different colour or shade (the door or the front fenders might look shiny compared to the rest of the car). This means you can be very sure that the car has been in an accident before. Due to the above reasons and due to a few other, most used car owners polish their cars before displaying them in the market. They make them look attractive and mask by polishing them into shiny new cars. Though a polished body looks all glossy and shiny, the polishing process has its flaws as some parts of the body turn out be lighter or darker and with a close thorough examination one could even find traces of polishing paste on the surface. If the body was polished using a polishing machine, you can see characteristic circular traces while observing the surface from an angle. As we all know that during the painting process all the glasses are removed from the body. However, most people do not do this, instead simply cover the glass with paper. But the manufacturer of the car inserts the glasses into the body panels after the paintwork, at the factory. Knowing these little details can help you to determine if the car body was repainted or not. To find this out slightly raise the rubber beading(gasket) around the windshield or rear window, if the colour underneath is different from the rest of the body, then this car was repainted. If the new paint is visible on the seals already, you don't need any further clarification as you can see the car has been repainted in a unique and obvious way. If in doubt, another check can be made by opening the trunk and comparing the colour inside with the outside. It is recommended to raise the trunk floor mats and assess the condition of the paint under them. Remember that these places on a used car would never look perfect and there would be at least some small chips, rust, scratches or other traces of the trunk being used. A similar test can be performed in the engine compartment by carefully examining the cup holding the battery, or the frame, etc. When inspecting the body, do not forget to check the condition of the front side of the doors as there is often rust in this zone.

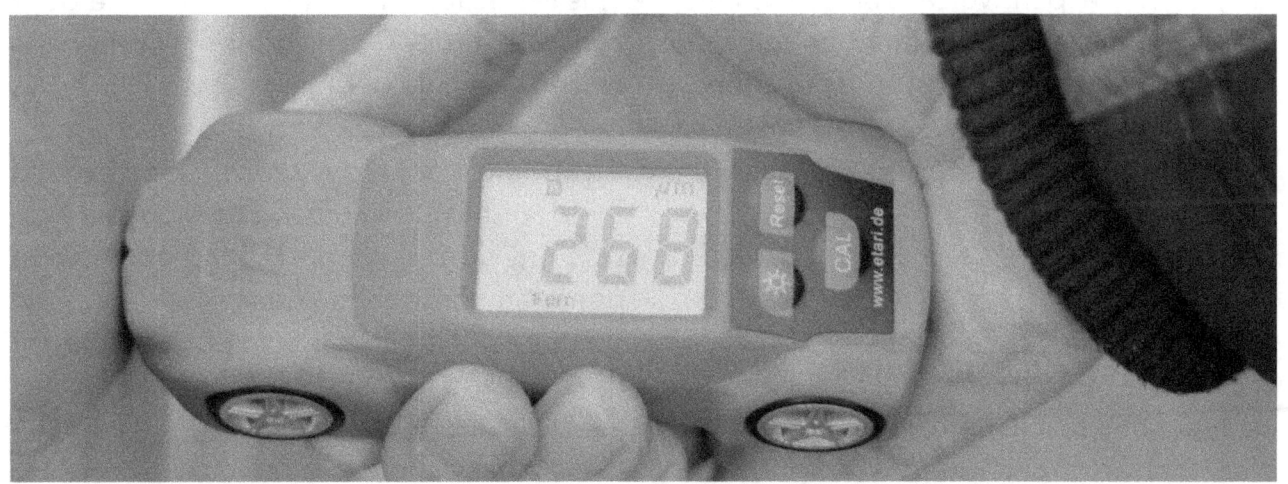

If you see characteristic spots with a yellowish tinge on a polished or repainted body, it means most likely, the metal under the paint began to rust, which is an indication of a poorly maintained car and you are being misled. Tip: When inspecting the vehicle, especially the body, be sure to check the condition of the following elements such as sills, side members, wheel housings, underbody, corners of the fenders, around headlights and under the moldings. In addition to these, many interesting things can be found under the luggage or cabin floor mats. If there is fresh paste somewhere on the body, you should be careful as many dishonest sellers use this to cover up signs of corrosion. In any case, the paste-coated surface should look intact, without any signs of corrosion such as blistering or cracking. A large amount of paste under the fenders is a bad sign as traces of low-quality welding are masked using this method. To identify the true age of the car, observing the chrome parts help. This is because over time chrome parts lose their shiny effect (they can also show stains of corrosion, which cannot be removed so easily or fade off, etc.) even in expensive cars. Since the chrome plating process and other operations to restore chrome parts cost a lot of money, sellers prefer not to do it and this helps us to determine the actual age of the car.

TIP

An ordinary magnet can help assess the true state of the car body. If it doesn't stick on the body at someplace then most likely, there is a thick layer of putty here which indicates that the car has been in a serious accident before. It is recommended to place the magnet near to the ends of the doors, wheel arches and on the curves of the fenders. If it doesn't stick, then the metal in these places is badly damaged by corrosion. Looking around the body, do not forget to check the doors. Opening and closing doors should be done easily and smoothly without any serious effort. Pay special attention to whether there are any traces of scratches due to grinding of the doors on the coating near the door and door frame. If you find such defects then you can conclude that either the body geometry is broken (usually this happens after a serious accident), or the doors are poorly lined up due to metal corrosion or weakening of the hinges. In any case, purchasing this vehicle can hardly be considered a good idea. Another important point that should not be forgotten when inspecting a car is to check the condition of the locks and this includes the safety locks within the latches. Remember that a malfunction of the locks can cause the doors to open abruptly when the vehicle is moving, which may lead to a serious accident. This is because, with a faulty lock, the door can remain open even after the owner has closed the car and might seem like it has been locked. Check for rust on the hinges. If you find it then the rubber seals around the door has worn off and they do not provide any tight sealing enabling moisture to penetrate through the gaps. This defect can quickly lead to corrosion at the bottom of the door and damage the window regulator as well. A lot of interesting things about the true state of the car be told by examining the hood. Try to open and close it three to five times to make sure that the mechanisms are working. Check that it sits perfectly once you close it and also pay attention to the dimensions between the hood and the adjacent body parts on the right and left as they should be the same. Otherwise, it can be said with absolute confidence, the car was involved in a traffic accident.

ATTENTION!

Pay special attention while checking the condition of the hood lock. There have been many cases where the hood opened while the car was moving and led to serious accidents. If the car is in good condition as a whole but the only issue is the hood lock, then this isn't a reason to refuse the car. This is far from being a serious damage and can be easily fixed either on your own or at a service station at a cheap price.

HOW TO DETERMINE THE TRUE AGE OF THE CAR FROM EXAMINING THE INTERIOR?

After completing the external inspection of the body, you can proceed to the next stage, which is inspecting the cabin. Apart from other aspects, inspecting the cabin can determine the accurate mileage of the car. It's not a secret that many mechanics can easily change the odometer readings (of course, on a smaller scale). Firstly, visually assess the condition of the steering wheel, upholstery, the doors and the seats. For example, if you find no patterns on the rim of the steering wheel, it indicates extensive use over the years and the mileage should be high.

Heavy seat wear may indicate that the car was used as a taxi. Damaged seats clearly indicate that the car was operated mercilessly over a long time.

TIP

Never buy a car which was used as a taxi. Such cars are in a terrible technical condition and can only be sold off as driving school cars. Coupled with long years of intensive hard usage, in most cases taxi cars are poorly maintained (the parts replaced according to the "less expensive" principle, poor quality and cheap oil is used which gets worse over time, etc.). As much as possible, try to avoid buying a car that did not belong to an individual, but to a legal entity such as a corporate. It is no secret that people maintain their own cars better than those which are for temporary collective use. The presence of cracks on the steering wheel can indicate that the car was involved previously in a serious traffic accident.

This is due to the fact that during an accident the driver usually "bumps" on the steering wheel with his chest, which leads to some cracks on the steering wheel. When examining the interior, don't forget to pay attention to even little details such as various stains, stale upholstery, holes burned by cigarettes, etc. All this can significantly reduce the cost of the car. In addition, these signs clearly indicate the intensity of vehicle operation. Also look at the condition of the rubber pads over the pedals and if they have strong abrasions (Fig. 2.4), we can conclude that this car has done a considerable mileage.

If the car which looks obviously worn out has good rubber covers on its pedals, then most likely the seller wants to deceive you and prove that the condition of the car is better than what it is actually in reality. Don't forget to notice any strong deterioration of the driving mat in the area of the gas pedal (sometimes they could even have holes) as this may also show that the car has done really high mileage.

Heavy seat wear may indicate that the car was used as a taxi. Damaged seats clearly indicate that the car was operated mercilessly over a long time.

TIP

Anyway, it is not a good idea to buy a car which in a condition that is clearly contradictory to the reading on the odometer. Also when inspecting a used car, make sure that the seat mechanisms are working, check the working of the horn and lights, windshield wipers, its washers and dashboard/cabin lighting. Check if the fan is functioning normally (you shouldn't be getting any strong smell as you turn the fan and heater on).

ATTENTION - THE WHEELS!

Some inexperienced buyers do not pay close attention to the condition of the wheels (fig. 2.5) during the external examination of the "used" car, and make a serious mistake. The wheels allow us to learn a lot about the true condition of the car. For example, uneven wear of tyres can indicate serious malfunctions in the suspension, steering column and it can also mean the body geometry is misaligned.

Experienced drivers are well aware that cars eat rubber. This means that along one edge (it can be both external and internal hem) the tire is entirely worn off. This may indicate that the car has a poor wheel alignment, but this is not the worst possibility. It's also a symptom indicating a car's really bad past such as a serious accident.

Be sure to evaluate the condition of the wheels, it's bad if they are rusted and very bad when there are dents, etchings and other damages. Another alarming indication is the presence of a bump on the sides of the tires. It indicates damage to the steel wire inside the rubber. Operating a car with such tires is dangerous and illegal (this is

stated in the Basic Provisions for admission of operating the means of transportation). It is a complicated situation as the „bumps" are very difficult to find visually. But they could be detected while driving the car, especially by analysing the front wheels. This is because the steering wheel will tend to rotate or pull towards one direction and when you press the brake pedal there will be vibrations. As for the „bumps" on the rear wheels, they are difficult to detect and only an experienced driver can do it (he can feel it from little swivels while driving).

The poor condition of the wheels may indicate that the machine has been operating in worse conditions and was not maintained properly which means it's doubtful the suspension and other parts will be in good condition.

PROBLEMS WITH SUSPENSION AND TRANSMISSION

Suspension and transmission are very crucial parts of the car, inspecting which would let you to find a lot of interesting things about the condition of the car with a high degree of accuracy, which is often completely different from what the seller tells you. In this section, we will talk about how to assess the condition of the suspension and transmission at the place of sale or the market. Checking the suspension usually starts with inspecting the shock absorbers. The most well-known technique is applying sharp and strong force on the top of the cor-

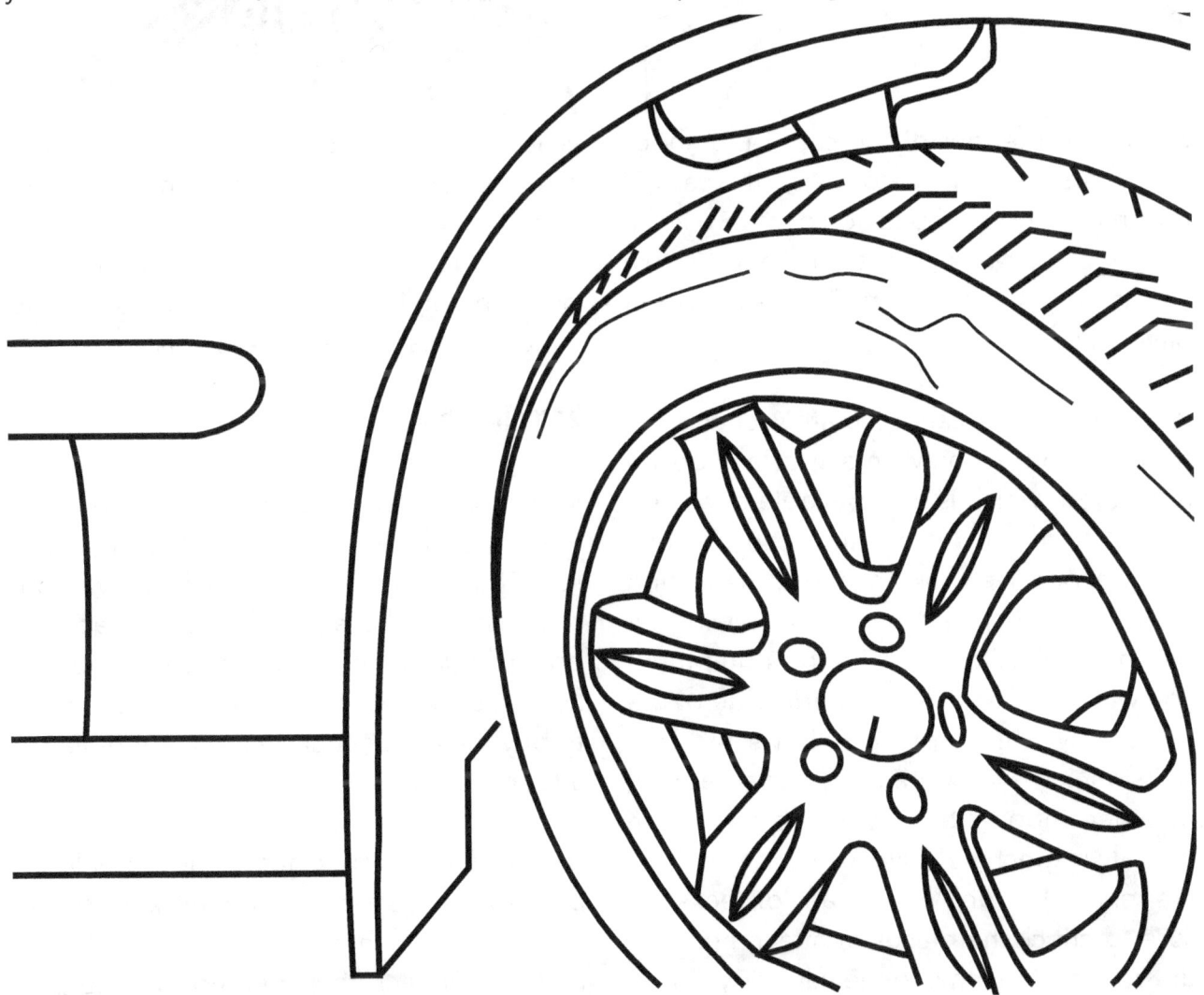

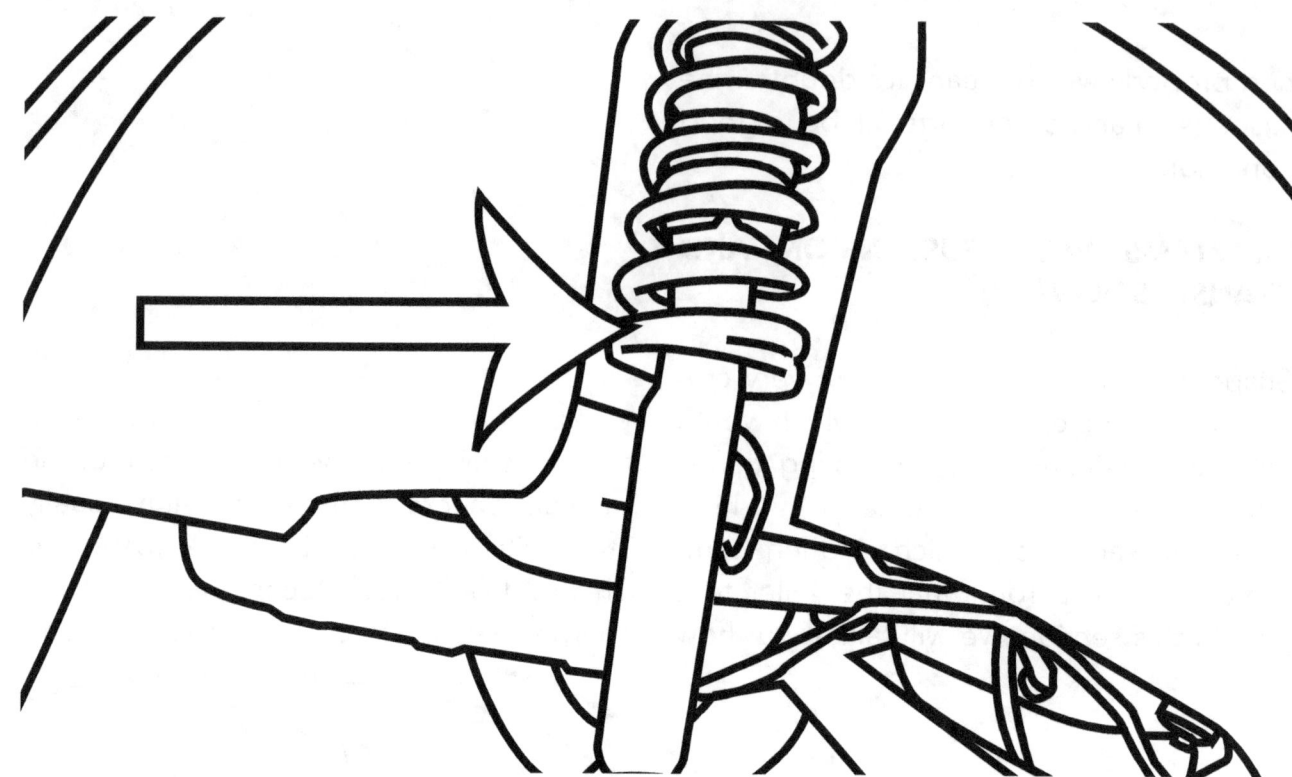

responding fender of the car. If the shock absorbers are in good condition the car will return to its original position of after oscillating once or twice. If possible, bend down and inspect the dampers. They should not have any traces of leaks.

Remember, you can accurately check the shock absorbers only on a specially designed stand, which is not available at all service stations. This procedure can be relatively expensive. Talking about inspecting the car it is necessary to have a viewing pit or a hydraulic lift. Only with this equipment you will be able to see the underside of the car clearly which not only shows the suspension, but also some body panels as well as other important components of the car. Visually inspect the drive axle and gearbox, they should be dry, without any oil leaks. Be sure to check the places of attachments of the suspension arms to the car body. There

is an important fact here because, if the longitudinal beam was a bit deformed, it is not straightened (this is quite laborious) back in many cases. In this types of case, the suspension arm is adjusted to fit straight using various types of bushes and washers.

ATTENTION!

While buying any used car, you need to carefully check the condition of the brake hoses. They must be fitted perfectly without any rust and there should be no traces of brake fluid on them. If you find that the brake hoses are in poor condition, either refuse to buy the car or change them immediately after purchase.

And in general, special attention should be paid to the inspection and working of the brake system, since this is a matter of safety. Firstly, check the condition of the parking brake. You can do this by engaging the handbrake completely, shifting the first gear and gently releasing the clutch pedal. If the parking brake functions normally, the motor will stall. However, if the car moves while the handbrake is engaged then this is a serious issue. In this situation, you should check the usual brakes by using the following method. Use your left foot to engage the clutch pedal and position your right foot diagonally so that you have your toe on the brake pedal and the heel over the gas pedal. Start the engine, engage the first gear and try to move the car by pressing the gas and the brake simultaneously while disengaging the clutch. If the car starts to roll and drive, it means that the brakes require repair or at least some maintenance. The most common causes of such brake defects are the presence of air in the brake system, wear of brake pads, discs, etc. The next important point is assessing the efficiency of the clutch mechanism. Despite the fact that many car owners try to fool buyers with lies such as "you can still drive on this clutch", "Yes, I just changed the clutch a couple of months ago,", you shouldn't be naive to believe them. Most sellers do not recognize when the clutch completely dies. To find this out, start the engine and press the clutch pedal and listen to how it works. If you hear a quiet screeching or hissing sound then, most likely, the thrust bearing of the clutch is almost destroyed. Be sure to try shifting gears while the engine is running (of course, using the clutch pedal) and if you feel that they shift with difficulty, then it means the clutch does not disengage completely (especially if there is a characteristic metal grin-

ding sound), which is an indication of a malfunctioning clutch. Here is another simple trick to determine the accurate condition of the clutch. Start the engine, engage the first gear, start rolling and pay attention to the position of the pedal where the car started to roll. If the car started to move immediately as soon as you let go of the pedal, the clutch works normally. But if the car started to move only after releasing the pedal until the end then it means that the clutch is worn out and it needs to be repaired or changed in the near future. And here is a tip for identifying hidden defects in transmission. Start the engine, engage the reverse gear, drive a few meters in reverse (just a few meters), then press the clutch, shift into first gear and start moving forward. While doing this if you hear a characteristic clicking sound from the drive axle or the gearbox, then the transmission is most likely worn out (the louder the sound then higher the wear).

You can independently check the play of the axles by rotating them and moving them back and forth. A slight gap is fine but not a lot. If in doubt, check the car while turning. If a certain crunch is heard during turning, it means that you need to change the joint that transmits equal angular velocities (SHRUS) as soon as possible from the side from which this crunch is heard. The is not cheap as the CV (constant velocity) joint itself is expensive and the procedure for replacing it costs extra money. But to be honest, if you come across a decent car which has just a faulty CV joint then you should definitely refuse to buy it.

WHAT YOU NEED TO KNOW WHILE INSPECTING THE ENGINE

The engine is the heart of any car. So the sellers usually wash the engine to make it look shiny, add a lot of additives and special oils to make it run noiseless and smoke free (which lasts only for a brief period) to increase the marketability of the vehicle and fool the buyer.

Before you start the engine, you should conduct a visual inspection. The engine should not have traces of leaks, dirt, etc. Another important point is the appearance and condition of the bolts, if they have any broken edges or other mechanical damage. This may indicate that the engine is far from being the good one. This raises a logical question about the quality of repair and in general, about the condition of the engine. To be more precise the presence of oil stains on the engine does not always indicate serious malfunctions. In this cases, it is necessary to understand what are the nature of these spots and why they appeared in the first place. For example, if there are traces of motor oil around the oil filler neck then there is no big reason to be concerned as most likely someone just inattentively and carelessly poured oil into the engine which flowed out and caused stains.

REMEMBER THIS

The presence of traces of gasoline/petrol on the engine is an extremely alarming and dangerous situation. It's better to avoid and not purchase these cars. Note: this can cause the vehicle to ignite anytime without warning and lead to a catastrophic accident.

After a visual assessment of the engine, it is recommended to inspect it again after the test drive. As we know from earlier the seller might have washed the car and engine to make them look shiny and appealing, but after a test drive any signs of malfunction on the engine will be evident (especially if it is in poor condition), though it may have a good external appearance. For example, the same gasoline or oil leaks, traces of antifreeze/coolant, etc. can occur again. Note that, traces of coolant leak always don't indicate serious problems with the engine. For instance, if the antifreeze oozes out because of a radiator leakage or due to overflow of reservoir, there is nothing serious to worry about. But the presence of any traces of antifreeze on the hose connecting the radiator and the expansion tank, or the presence of cracks on this hose may indicate the engine is old. Every buyer should remember that the state of the oil inside the engine can reveal many interesting things about the condition of the engine. However, used cars sellers are well aware of this, so in most cases they change the oil before selling (of course, they fill it up with the cheapest and lowest quality oil, which is why after buying a used car, you should always change the oil and filters).

NOTE

Often, before heading to the car market the car owners pour too much thick motor oil into the worn out engine of their car. This simple technique sometimes effectively muffles down the noise of the almost dead motor. But an experienced driver can easily determine the viscosity of engine oil by touch.

Sometimes used cars have a punctured cylinder head gasket. This causes the coolant to enter and mix with the engine oil. When visually inspected, this fault is imperceptible and highly unnoticeable. But the issue here is, removal of the gasket (and replacement) will cost a lot of money as it requires a lot of time and labour. To check the condition of the gasket just start the engine and before it reaches the operating temperature and look into the coolant expansion tank. If you see bubbles in this tank, it is possible that the cylinder head gasket is damaged. Such a test does not absolutely guarantee the above situation, but the probability of a defect is high. Another important symptom indicating a gasket malfunction is the presence of oil stains on the radiator. An important point when visually inspecting the motor is checking the condition of ignition system wires. If they have lost their elasticity, either completely or partially and show signs of oxidation or cracks then this is a bad sign, especially in wet weather (the engine starts up with difficulty and works intermittently) it could cause lots of troubles. If everything looks clean, but there are traces of rust in some places and paint is scrubbed of at a few places, this is a probable indication that the engine has been cleanly washed recently before the car was about to be sold. If the owner of the car had constantly kept his engine clean such signs will not be visible most likely. If a visual check of the engine condition did not cause you any suspicions, then it is safe to buy. A fully serviceable engine, even in a cold state, should start without problems and work stably.

REMEMBER THIS

If the Engine of the car is in good condition, then a poor start may be due to its old age and insufficient battery power. Possible causes of this may be problems with ignition or malfunctioning of the spark plugs. To mention here, one of the most common tricks of sellers is that they will deliberately replace an old, outdated battery into their car while selling it. However, if there are no complaints and the battery seems to work fine there is no reason to refuse purchasing it. The same applies for sparkplugs too. But if the car ignition system works badly then it is better not to get involved with the vehicle. Therefore, sellers often try to fool the buyers, arguing that engine starts poorly because of just a bad battery. It is easy to check the veracity of such statements by trying to start the engine from a different battery (for example, connecting to the battery from another car). If you are buying a car with a diesel engine, then remember that the thick black smoke from the exhaust pipe immediately after starting the engine is not a sign of any serious problems. This is typical, for cars equipped with a diesel engine. This fact could also be used by dishonest sellers to con you. Because the thick black smoke should quickly disappear after you start the engine and run it for a sometime and if this does not happen, then something is wrong with the machine. Many owners often try to convince the buyer by saying that „everything is in perfect condition; this is a common phenomenon in diesel engines". By the way, such a sign of constant black smoke may indicate serious problems in the injection system. If you see thick black smoke out of the exhaust and the engine produces a characteristic loud knock at the same time, it is likely that the injectors are faulty. If you are checking a gasoline engine, after you start it, let it run for a few minutes until it reaches its operating temperature. Then, open the hood and listen carefully whether the motor "trots", this is a sign of knock if you are able to hear to a weird metallic clunky sound from inside. Then look at the exhaust gases, the smoke coming out of the exhaust should not be too thick and also black or blue. If you notice that the car spews out exhaust smoke having a bright blue colour, then there is a very high probability that the engine is malfunctioning. Most often this sign indicates the wear of the piston rings, in which case, the engine oil enters the combustion chamber leading to the appearance of blue smoke. In such vehicles, in addition to the increased CO content in the exhaust gases (such a car will undergo a technical inspection), this also leads to an increased consumption of engine oil. This type of situation is very unfavourable as the motor needs a major overhaul and this is a very expensive process. It's best to refuse buying such a car, no matter how much the seller tries to convince you otherwise. As for cars equipped with diesel engines, many people don't suspect anything if exhaust gases are almost colourless. If the exhaust gas is in a greyish or white colour, then it means the coolant is entering into the cylinders. As we saw earlier, this can occur due to damage of the head gasket. Sellers of diesel cars often try to explain that this exhaust smoke colour is common in a diesel engine, which is criticisable. And also the greyish or white colour of exhaust gases may be due to improper spraying of the nozzle or incomplete combustion of diesel oil due to improper injection, as well as other serious problems.

If the seller who has a huge mouth, claims that the diesel engine of his car is "almost new" then do a little test. Ask him to start the engine, you open the hood and disconnect the ventilation hose (it is easy to identify it as it's thick) that goes from the oil casing or from the cylinder block to the air filter or intake manifold. See if any thick smoke comes out of it (like fog). If you see thick smoke, then you might need to check if the engine oil has penetrated into the air filter and intake manifold. If this is the case, it means that he is trying to deceive you and engine is worn out.

TYPICAL TRICKS AND TRICKS OF CAR DEALERS

Previously, we have emphasised that the main goal of any car dealer is to sell the car as quickly as possible and with the highest profit. Keeping this in mind the buyer must be careful to avoid being swindled. The seller always welcomes the potential buyer so cordially and kindly, as if he was waiting to meet him his entire life. This, of course is a flattering behaviour and immediately after receiving money from you, he will lose interest in you and disappear in an instant. And even if he gives you his phone number (in case you need to clarify something with questions such as "how to turn the air conditioner on?", or "how to fill up gasoline?", "where is the pump?", etc.) on your request, it is quite unlikely that you will ever call him later or most probably he will not attend the call. Many of those who professionally earn their living by selling cars, are very talkative and sometimes the buyer does not have any time to even talk a couple of phrases during the conversation. This is a trick of the seller to keep constantly talking to the client triggering his desires and preventing him from orienting himself in the situation, thus making him like the car without inspecting it properly. Professional marketing people who sell cars in the car market work in pairs as it is easier to sell cars this way. They agree cooperating in advance, but while in the market they sell their cars not side by side, but at a considerable distance from each other (might be 30 cars between them). Usually the sellers working together sell cars of brands, models and class which are different from each other so that the same buyer does not accidentally wander into both of them at the market. When one of them is approached by a client who clearly shows interest in the car, this seller invites his partner on the mobile phone. He quickly arrives at the scene and plays the role of another buyer, telling things like he liked the car very much, the technical condition is very good without any complaints, in short the car is just perfect and so on. At the earshot of the buyer the partner makes an act of agreeing to buy the car and usually tell something like "I want to buy this car but I don't have the cash right now. Could you please hold the deal for half hour until I get the money? "to which the seller responds "I'm sorry but I cannot guarantee anything as I have other buyers who might be interested before you get the money". The partner acts like he is frustrated, looking very sad and with impatience and begins to look at the buyer, gazing at him hoping he would make a hurried decision. In my experience, this is a fairly effective method as people often agree to buy a car "in excellent condition", which already has a queue of competitors lined up or starting to line up. And the seller provides similar help to his partner in the sale of his car. Such partnership is not between two buyers but other people could also team up with the seller. In many markets there are teams of such people who operate successfully for a long time for specified percentage of the car price of the car they help to sell. They are capable of playing really good ideas before buyers that even a psychologically stable person could be fooled and would buy the car.

One should always remember that an experienced seller is able to determine the psychological state of a potential client by just looking at him and behaves accordingly after analysing him. If finds a man who cannot sleep by the mere thought that he does not have a car and is determined to buy it today or as soon as possible, he can easily "drag him into buying his car", having just said a couple of sweet welcoming words. Such a person has become filled with impatience with the idea of owning a car for a long time now and just a little nudge is enough for him make a decision without a thought. Such clients, usually almost never look at the car, making a vague assessment of the vehicle and rush to pay as soon as possible, fuelling their impatience to get behind the wheel of their own car. Similarly, the seller recognizes a serious, unresponsive customer who isn't overemotional and who carefully inspects the car. In this case, the seller is aware that being talkative is meaningless and such buyer will not pay any attention to these tricks, in the worst case he will simply wave his hand and go to watch another car whose seller who isn't so annoying. If a doubting and a quite self-confident person approaches, the seller will not talk too much and praise the car too much as this may scare the customer away. In this situation, the seller will explain the advantages of buying his car, with a soft tone and with a confident string of arguments (albeit, in fact, arguments may turn out to be dubious). Sometimes sellers successfully manage the following trick. If the buyer examines the car for a long time without making any decision to purchase, they can say something like "This car is not for sale, someone already booked it and went to get the money". After this if the buyer leaves it's unfortunate and the seller might lose a customer. But if he doesn't leave and looks at the car with regret that it slipped out of his hands, then the salesman, might tell him something with a doubtful tone that he is ready to sell the car to the him but for a higher price since he already promised it to someone else and is reconsidering the decision for him. If the client does not mind and purchases it the call will be sold as if it's brand new! Depending on how the situation builds up (for example, if the client does not have money with him, or has doubts), the salesman can act accordingly. In this case, he can sell the car at the stated price, without any additional tricks and would claim that "only out of respect for such a good person" he accepts the promise to receive the money later and sells the car. Usually a government ID or passport is left as a deposit or some advance is paid (100-200 euro) which will subsequently be included in the final price of the car. In general, sellers use a wide number of psychological tricks. Another trick is as the buyer is looking at the car he likes the seller strikes a spontaneous conversation with seller standing at the next stall during which they anxiously discuss about the increase in customs duties on foreign cars which are coming out in the near future. The conversation is loud enough for the buyer to listen to and is filled with phrases like "yes there is an increase and I heard about it", "the cars started selling off very quickly before the introduction of new duties and did you hear that Volodya managed to sell two cars yesterday," etc. Often, the client bites the bait and decides to immediately buy the car assuming that he prices may go up in future.

HOW THEY CHEAT YOU IN CAR SHOW-ROOMS

It is a common belief that while buying a car from the showroom or an authorized dealer, people are 100% insured against any deception and unpleasant experiences. But in reality, things are different and sometimes end up entirely opposite to what you expected. Despite the high prices at the showrooms when compared to the market, buyers can be deceived by showrooms too and sometimes the size of fraud is really depressing. And it is especially unpleasant when you end up being cheated at the place where you least expected.

REMEMBER THIS

You can be deceived in almost any Russian motor show, even the most "cool" and prestigious ones. Consequently, you should trust a car dealership or showroom no more than you would trust an individual seller or an advertisement.

Usually car dealerships use a technique to subtly get more money from the customers. When the customer looks at the cars and chooses one, he would be informed that the car has already been booked by some other person just before he can purchase it. The customer would have no idea that a situation like this might arise at the last moment as, once he enters the showroom, he would get a warm welcome, treated with great hospitality, would be politely shown around and explained about all the vehicles etc. As soon as he makes a decision to buy a car, a bunch of paperwork begins immediately which will increase his excitement and hopes of driving the car home. And suddenly he would be informed there has been an unforeseen situation and they would explain him that someone has already booked the car. And coincidentally according to the showroom salesmen that car would be the last one available and this would mean that our guy needs to wait for some time which is usually a few months. Such an unpleasant situation can spoil the mood of any person who had lots of anticipation. As he wanted to purchase a car as soon as possible, had the money ready, but now things turned out unfortunate and he has to wait for a long time. And just at this moment someone from among the dealership salesmen would come forward to resolve the issue. He would assure the customer that he could negotiate with the competitor and could convince him to postpone the purchase. It sounds good but the catch is, he would tell that it would cost extra as the other customer is giving up his booking. Most buyers are fooled by such ploys and pay extra just to quickly own a car. In reality there was no order on this car and it was all a plot to swindle the customer.

OTHER TRICKS USED IN CAR DEALERSHIPS

Purchasing a car at a showroom or dealership has many subtle things which you should be aware of. You keep in mind that the workers in these places are experienced specialists and their income almost directly depend on the sales. Therefore, there are more interested in selling the car to the client no matter what, than caring about the client's needs, requirements and interests. Female buyers are in a greater disadvantage as they have relatively less idea about cars compared to men. Using this as an advantage, they could be persuaded to buy something completely different from what they expected. by the time they realize the fact they wanted to buy a completely different car it will be too late as they would have already bought a different car and drove it for a few miles.

TIP

When Choosing a new car and getting familiar with its basic features and characteristics, try to use several sources of information, including the official web page of the brand or car you are buying. Please note that the data and specifications of the car may not be completely accurate (especially about equipment and technical characteristics) in some websites. This also applies to online promotional advertisements or the dealer website.

When purchasing a car at a dealership or showroom, don't forget that the employees or salesmen may have poor knowledge of the cars and would not have complete knowledge about all the technical specifications, features and other information. It could be even worse, when the salesmen try to deceive you on purpose giving you false information. Remember that their target is to first sell all the cars which are in the dealership inventory sitting unsold. To sell these cars as soon as possible these salesmen would go all out to convince you that this is the perfect car of your dreams. Many car dealership employees are not honest with their information and can mislead you by modifying some data. They could "increase" luggage space, "shorten" the car's acceleration time from 0-100kmph, "reduce" the fuel consumption, etc. from the actual numbers. Another common situation is when car dealerships persuade a client to buy a car which he didn't want to buy in the first place. For example, they could offer you a car with better performance and higher cost. In such cases, you need to firmly and confidently tell them that you need a car with the specific characteristics and features that you already mentioned.

If you are purchasing a car at the beginning of the year, be sure to check the year of its release. This is because at the beginning of the year, models produced the previous year are usually sold off (although showrooms and car dealers deny this fact). New cars are being introduced by brands every year and the cars from previous year can get cheaper even if it's brand new.

If the car you like is not available in the showroom at the moment and you agree to get it delivered after a few days by placing an order, do not forget to check the delivery time on contract. Salesmen in some showrooms and car dealerships can be cunning as they verbally agree with you on the delivery time for 2 months. let's say but would mention 75 days on the contract. Obviously, that's a very noticeable difference.

It is also necessary to stipulate the responsibility borne by the showroom or the dealership, if they don't fulfil their obligations mentioned in the contract. For example, according to the contract you should have received the car in 2 months but it took 70 days for you to receive the vehicle then the showroom or car dealership should pay you penalties for violation of delivery terms (you can specify in the contract the everyday delay will cost the seller about 0.5% of the price of the car).

In the same way, you can insist on payment of penalties in an event that the car was not delivered at all due to uncontrollable circumstances despite the fact that you prepaid the money in advance. In these situations, employees in showrooms and car dealers usually give you an innocent apology which is something like: "We apologize, we could not get the car for you, so you can get your prepayment back at any convenient time". But at the same time, they don't mention the fact that they (car dealership) have been using the funds of the buyer. Therefore, a logical solution to the issue should look like this: The prepayment paid in full plus also the amount of penalties for non-fulfilment of the terms of the contract of sale and purchase, and besides this also the interest for using the prepayment funds is returned to the buyer.

AFTER THE PURCHASE

» After becoming a happy owner of a used car, registering and sharing your purchase with your friends, perform very few simple, but extremely necessary procedures:

» Replace the oil in the engine, as well as other components and consumables. For example, the oil in the gearbox, yes you need to change the gearbox oil every 60,000 km. Many people think that only automatic gearboxes need frequent oil change, manual gearboxes also require oil change though the manufactures often don't mention this fact.

» Replace the antifreeze. It is not always clear in how the engine was operated previously, how often the coolant was changed and how efficient the radiator is.

» Drain the brake fluid and refill the system. This will save you from trouble if the brake fluid had absorbed a lot of moisture previously which can lead to increased viscosity;

» Visit the tyre balancing centre. Most probably the previous owner did not check the alignment and the wheels are misaligned.

» Change the air filter and cabin filters. Starving the engine of oxygen is destructive. The cabin filter provides more comfortable conditions inside the car with the windows closed;

» Replace the timing belt and pulleys. It is not always clear how much the previous owner used the car on them without replacement. They could be the factory ones or aftermarket parts;

» For cars older than 7 years, it's better to replace crankshaft oil seals, gearbox seals etc.

Of course, you don't have to change all these at the same time because some changes like replacing the timing belt is not cheap. Demand the previous owner a service book or some kind of confirmation that the timing belt was replaced before 20,000 km.

VERIFICATION OF DOCUMENTS

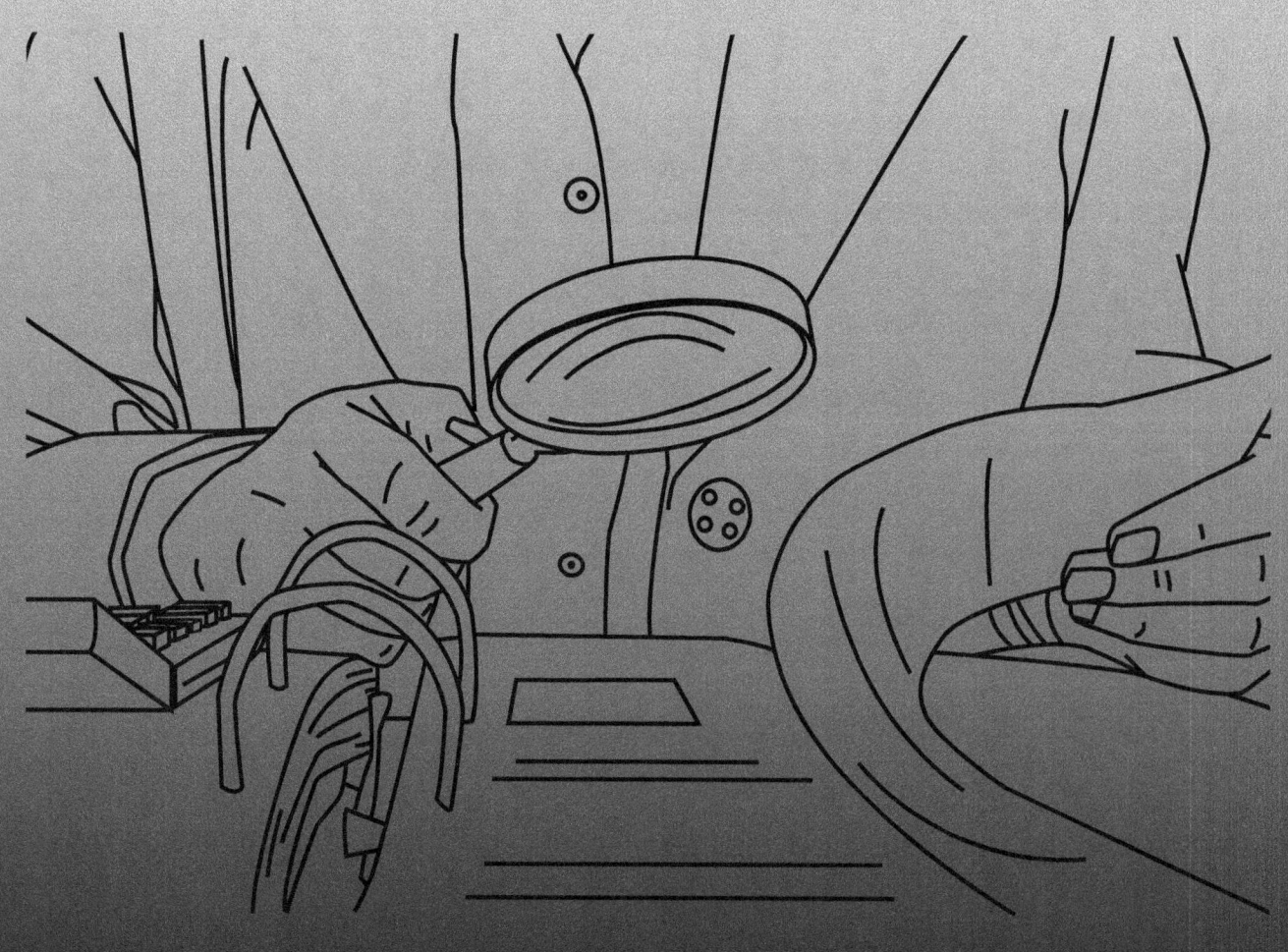

WHAT IS THE CAR'S VIN-CODE AND WHY IS IT NEEDED?

Surprisingly, many motorists do not know about the existence of Vin-code of the car, though this has a lot of useful and valuable information. The vin-code is a unique identification number of the car, assigned by manufacturer when the car is rolled out of the conveyer belt. Vin-code is assigned to all cars released after 1980. The format of the car's vin-codes is described in the ISO 3779 standard, which was adopted in the USA and Canada from 1977. This standard is followed by the vast majority of European manufacturers.

TIP

When buying a car, it is recommended to check its Vin-code, especially if this car is expensive, rare or an exclusive one. Owners can change the code numbers which is really an inappropriate thing to do. Suspicions can arise by just visually inspecting the vin-code. Pay special attention to characters that are relatively easy to correct (for example, the number 3 is corrected by 8, and the figure 5 - by 6, etc.).

Note that when checking the vin-code, you can not only evaluate its correctness, but also check other characteristics of the vehicle (list of vehicle options, etc.).

Replaced vehicle passport should arise serious suspicion.

It should be suspicious of those cars that have a replaced vehicle passport which could be issued to replace the previously lost one. Especially in the case of new car models. There is a high probability that the vehicle passport was not lost, but could have been through some underground deals related to theft and fraud.

TIP

If a car with a replaced passport is sold to you at a low price, refuse to buy it without a shadow of a doubt.

THE PSYCHOLOGY OF THE SELLER, OR HOW TO CHEAT BUYERS

Why do sellers try to cheat while selling their car? Simply putting, the car is the second most expensive purchase in the life of an average person. Therefore, people try to sell it for a good value by any means possible even if it is a pile of trash.

It's no secret that every car seller (whether he is selling a new or used one) has a whole range of a wide variety of psychological techniques in his arsenal aimed at convincing the buyer to purchase his car. However, this is just half the trouble, because when you see the situation from a humane point of view, sellers are humans too who sell their cars to solve their monetary problems in life. The main issue here is to not become victims of fraudsters. It's no secret that the car market is one of the most opportunistic and favourite places for fraudsters, robbers, con men etc. of different varieties.

In this chapter, you will learn about the techniques and tricks used by car dealers, as well as how to avoid fraud and deception when buying your four-wheeled friend.

MAIN FEATURES OF THE PSYCHOLOGY OF SELLERS

As I noted above, the target of the seller is selling the car as soon as possible at the most profitable price. It's very rare to find sellers these days who don't try to attract customers with a beaming smile and meet him as if he had dreamt about this moment for a long time.

Any experienced seller can assess the psychological state of the buyer and depending on this, he conducts himself accordingly. For example, if the person before him is obsessed with the idea of buying a car as soon as possible, he could easily sell the car to him by just saying a couple of attractive and false phrases. Such a buyer is already in the trap, all it takes is just some sweet words and then everything will work out on its own. Such buyers usually do not conduct a serious inspection of the car, just limiting themselves to a visual assessment and are in a hurry to pay.

If a car is inspected by a serious person who soberly assesses the situation, an experienced seller immediately notices this and acts accordingly. He understands that in this case it's useless to polish the buyer with his words and will remain quite without paying him too much attention, and in the worst case the buyer may get suspicious.

If someone who is vacillating and not so self-confident approach the car, most likely, the seller will not enthusiastically praise his car as this may scare the buyer. In this case, the seller will explain benevolently and calmly, in a confident tone with arguments (another question is, what kind of arguments?). He will explain the expediency of buying this particular car and the correctness of the choice.

For unsure buyers, sellers apply the following psychological technique. If they see someone walking around the call for a long time showing interest it but haven't made up their mind, the seller might tell them something like "You know, someone already wanted to buy this car and has gone to get the money, therefore it is not for sale". If after these words, the buyer turns around and leaves, that's the end of the deal. But if he continues to look at the car with regret that he missed a profitable option which slipped of his hands, then the seller would tell him that if he is ready to buy the car on the spot he could sell it to him, but for a higher price than stated (because he is selling the car to him despite promising it to someone else, which is fake drama of course). If the buyer agrees, the car will be sold to him.

Well, if the buyer is still hesitating or does not have the money with him at the spot, the cunning seller can offer other options. For example, he can offer to sell the car at the stated price without a premium claiming "it's out of respect for such a good person who came to the market without money". He would agree to the settlement of the cash the next day however only under certain conditions (because he promised him the car and waits until tomorrow for the money). Usually, he gets some advance cash as a token of promise, which will subsequently be taken into account when the buyer pays for the car (usually 100-200 euros).

Often sellers (especially in the automotive markets) use simple psychological techniques to subtly encourage the seller to buy the car. For example, while the person is inspecting the car, the seller has a casual conversation with another seller nearby during which they actively discuss the "in-crease in prices from day to day, because these models are in huge demand at the moment". Also for example, if a client wanted to buy an SUV but still not sure about it, the cunning salesman tells him that winter is approaching and he already received three offers for the car since morning.

Often, the buyer is fooled by such conversations and decides to buy the car immediately to avoid a surge in prices.

PRICE NEGOTIATION

Yes, bargaining is very important and significant part of the process as the final price of the car depends on it. You could even bargain with the manager at an auto showroom while buying a new car.

HOW TO BARGAIN WHILE PURCHASING A USED CAR?

If the price of the car is too much for your budget and you find an inactive music player in the car, you could bargain with the seller based on that. You can opt it out and reduce the price of the car.

THINGS TO RELY ON WHILE BARGAINING:

The defects and faults in the car. The amount of the discount must be adequate enough to cover the repair costs.

Comfort issues such as smoked out roof, missing ashtray or other stuffs that come standard in the car that are missing. You can bargain a few thousand over here.

Exterior defects such as tarnished chrome elements, the missing brand logo and even the absence of a visor is a reason to bargain.

Interior defects such as damages in upholstery. These could reduce the cost significantly.

Costs of dry cleaning the interior even if you don't plan to do it.

THE ABSENCE OF THE SPARE KEYS

If the car doesn't have extras such as the spare tyre, speaker system in the rear, daytime running lights etc. You could bargain using them as a reason.

INSTRUMENTAL INSPECTION

If all the documents are in order and the exterior has no issues, now it's time for instrumental diagnosis at the service centre.

While going to the diagnosis centre request the seller to let you drive. He would be happy to give you a test drive. While driving, try testing the car specifically over bumps and potholes at different speeds and angles. If the car passes over them without any knocks or squeaky sounds, then the condition is ok. If you could hear cluttering, creaking and squeaking sounds then you might need to discuss with the seller about these suspension malfunction issues and how long they have been existing.

YOU SHOULD ALSO EVALUATE

The working of the engine. The acceleration should be smooth and stable without any jerks or noises. Gear shifts should be smooth and fast with the exception of 1st and reverse gears as they aren't equipped with synchronisers. Clutch operation should be ok. The clutch should engage neither at the very beginning not at the very end of the pedal positions. Otherwise, you risk the possibility of needing to service your clutch and gearbox as soon as you buy the car.

Feel free to use electrical equipment. See if the dashboard indicators light up when you switch on and off the headlamps and fog lamps. Check the stereo, air conditioner, heater and power windows.

YOUR EYE IS YOUR FRIEND

While examining the body do not rush immediately and go around poking the car with a magnet or testing the thickness gauge on the fenders. First just go around the car looking for dents, scratches, scruffs in paintwork and rust. Ask the seller to show you the defects that were discussed when you communicated by phone. Check the uniformity of tire wear as this is very important. Try to estimate the tread depth in each wheel. Also check the spare tyre. If the wear is clearly uneven, most likely there are problems with the suspension which could cost quite some money to repair. But in worst case it could be an indication of broken body geometry.

Ask the seller to start the engine and pay attention to the colour and texture of the exhaust. Also notice if the engine starts without any issues and runs smoothly.

There should be no blue, grey or white smoke from the exhaust pipe.

If you place your hand at the entrance of the exhaust pipe, you should feel a steady pulse of air. But if everything seems good do not rejoice too early. Modern additives can make it look like the engine works perfectly but it might be needing a major overhaul in reality.

SEARCH FOR RUST AND CORROSION

Even cars that are less than 5 years old could be a victim of corrosion and even bad deep rusts can form on them. Almost all cars suffer from this "disease". After a serious accident or a hit, they can begin to corrode in a damaged area. Especially problematic areas are the door sills or edges where corrosion usually begins. Assess their condition will a neodymium magnet, soft rag cloth and a mirror. Here's how to check:

Visually estimate the thickness, colour and uniformity of the door sill;

With the help of a mirror, inspect the lower and the back of the sills, paying attention the holes made for jacks (if there are any);

Place the cloth on the body element and use the magnet to see if it sticks. If it is not magnetic, the element is filled up. If you see

no damage to the door as well as on the upper part of the sill when the door is opened, but the magnet doesn't stick anywhere then this means, there had been corrosion previously and these places had been filled up to conceal it.

A rusted sill is a reason for a serious reduction in price, since it's a matter of vehicle safety. It reduces the safety of the car as the rigidity of the body is significantly lowered, which is especially dangerous in collisions and can cause serious injuries or even death.

The sill is a completely replaceable part, so if the seller is ready to reset the price, you can think about the purchase. In most cases, the seller is aware of the repair costs, so depending on his decision it can either dramatically reduce the price of the car or the inspection could end right there. The next step is to use the thickness gauge to check the thickness and condition of the paint. This Inspection is performed in the following order of elements:

» Roof;
» Front bumper;
» Central post;
» Internal racks of pillars;
» Doors and sills;
» Rear rack;
» Edges of the hood;
» Adjacent fenders, including the wheel arches and in the place where the headlamp is attached;
» Rear bumper;
» Trunk lid.

Pay careful attention to the interior of the trunk. In most cases, the repair marks are not particularly hidden around here and simply covered using mats, rugs etc. Take a look into the spare wheel space, as after serious accidents the damage traces will be visible here, since this place is usually not restored with great effort.

If there is no thickness gauge, you will have to trust your vision, however modern computerized paint mixing helps to achieve 99.9% colour matching which can make it almost impossible to tell the difference.

If you find any high-quality repainted body, it doesn't mean that you should avoid the car. It's even common in cars bought in showrooms to be repainted at places because of minor dents during transportation or while reversing the car into parking lot. The major question is, what is behind the paint?

When conducting the initial inspection, ask the seller questions such as, reason the car was repainted, why it wasn't mentioned when talking on the phone (if the seller tried to hide this fact).

It is important to inspect the condition the body even if the car's history report does not contain data on the car being involved in accidents. Because, issues due to many minor accidents are resolved on the spot, without the involvement of the police or insurance companies. Or there could have been minor hits like, the former owner could have accidentally drove into a pillar at the garage or scraped the garage door while reversing.

The bumper should be given special attention to and inspected for any cracks or chips of paint. Since the bumper takes most blows during collisions or when driving on snow, curbs and other obstacles. For a normally operated car paint damage on the bumper is acceptable. The same applies to the front edge of the bonnet, especially with cars older than 3 years. When driving at high speeds small pebbles thrown out from the wheels of the car in front could leave scratches on the edge of the bonnet.

Rejecting a car just because it has damaged paint on the bumper or the hood is not a good decision. In fact, this indicates the fact that the car hasn't been in any head on collisions previously. Because many clever sellers replace the hood and bumper, trying to hide the traces of any serious accidents.

REPLACING BODY PARTS OR THEIR DISMANTLING FOR THE PURPOSE OF PAINTING

Any serious accident requires body work. Sometimes the damaged body parts are tinkered, but in some cases it happens to be that the body part must be replaced. A need for replacement arises only when the impact forces is too high.

Usually, there is nothing wrong in replacing the body parts, especially if it was designed good. The problem is that, the second hand market often offers cars that have passed though restoration process after accidents and this leads to questioning the condition of the body geometry.

Buying a car with a „destroyed" body can lead to a large number of problems. It is certainly less safe, tires wear faster, and directional stability is poor making them pretty useless.

Most often, the replacement of body parts/panels in the past is indicated by the discrepancy between the width of the gaps between the body panels. No matter how carefully repairs are made, it is impossible to achieve the complete precision compared to the factory assembly. Therefore, it is important to note these areas and gaps carefully, as they give away many details.

TIP

Replaced fenders, doors, hood or sills does not mean that you have to abandon the purchase altogether. However, this is a valuable reason for bargaining!

To determine the traces of the dismantled and replaced body parts you need to:

Open the hood, trunk and inspect the mounting elements. If there are traces of spanner or tools on the bolts, this indicates that the dismantling took place earlier;

Try out the doors. They must open and close with the same effort;

Assess the condition of the hinge door hinges. There should be no misalignment or traces of impact;

Try pulling the plastic door panels from inside the car. Replaced door covers often do not exhibit a snug fit;

Inspect the seals. Their wear must match the age of the car and shouldn't look too new.

Many dealers hide dismantling traces by painting the fasteners in body colour. But they could never be able to achieve factory quality, so when you find a difference in colour tone of mounting elements specify it to the seller asking why the repair marks were hidden.

IDENTIFYING PROBLEMS WITH BODY GEOMETRY

Visual identification of body geometry problems with the car body is usually possible.

To do this you need to:

Evaluate the evenness of tire wear. If the geometry is broken, the wear will be uneven due to the improper camber / toe;

Estimate the width of the gaps between body panels. There should be no discrepancies and the width of the gaps should be the same everywhere;

Assess the condition of the main frame. They can be seen both in the engine compartment, and from the bottom. There should be no folds, bends and even tiny traces of welding on the side members;

Determine if there are traces of roof repair. This can be done using a thickness gauge or by inspecting the tightness of the doors. It's better to abandon the purchase if the roof is deformed. Such a car will have problems with regularly cracking of Glasses, leakage of seals during rain.

PAINT QUALITY ASSESSMENT

As we have said earlier, if the repainting is done with good quality then there is nothing wrong with the vehicle. Distinctive features of high-quality painting:

Matching colour and tone to other machine parts;

Lack of shagreens;

The absence of stains, bubbles and dust under the paintwork;

No traces of paint on seals and other parts/elements.

Remember that the technology of factory painting eliminates the thickness of the paint layer.

High quality repainting technology under proper conditions and good service will show no difference in colour and has a uniform tone all over. Therefore, do not believe in any stories the seller narrates to make an excuse for the poor paintwork. And the more excuses he makes, the less trust he deserves!

EVALUATION OF LUBRICATION SYSTEMS AND EXHAUST

After inspecting the body, proceed to the assessment of the engine compartment and exhaust. For this you need to:

Check the level of all fluids (antifreeze, brake fluid, etc.).

They should all be above the MIN mark. Pay special attention to the colour of antifreeze and it shouldn't have a rusty hue. If signs of rust are present, the engine most likely has overheated previously;

Remove and check the dipstick. There should not be any lumps of oil over it. And it should not smell like gasoline. There should only be a thin layer of oil on it;

Inspect the oil filler neck. There should be no sludge on the lid.

The presence of fresh leaks near the neck indicates that engine oil could have been changed recently. Since the car is being sold, nobody chooses to spend on changing the engine oil so we can conclude that the seller is hiding something;

Inspect the oil filter (usually it is visible);

Try to check for leaks of working fluids on the outer case of the internal combustion engine, in places such as connections of hoses and nozzles;

If the engine is washed and clean, it could indicate a good presale preparation or an attempt to hide some leaks.

If the weather is warm and there is no precipitation, you can assess the condition of the engine by exhaust:

Exhaust should not be coloured. White colour indicates the burning of the gasket because the antifreeze gets into the combustion chamber;

Grey colour indicates the wear of pistons or piston rings as oil gets into the combustion chamber; black means the mixture is extremely enriched with gasoline;

Exhaust pulsation should be smooth and stable.

OTHER THINGS YOU CAN EVALUATE BY YOURSELF:

I have mentioned here a few more things which you can evaluate yourself and that will help you decide on the purchase and its value:

» Steering play. With the engine turned off, shake the steering wheel. If the steering turns too much but the front wheels don't turn, then there is an issue with steering play;

» Parking brake. Must lock the wheels on the 3rd click;

» The lock buttons, remote trunk opening and fuel tank flap.

» For some reason, only a few people pay attention to check the performance of these elements:

» Working of key alarm. There are cases where an alarm remote is attached separately to the keychain although the car has no alarm system installed in it! When you ask the seller why it isn't working he may come up with an excuse that the battery ran out just a day ago and it works, please don't believe such tales.

» Adjustment of mirrors. We all know that the side mirrors are very vulnerable in a city environment. They are often cracked by close passing vehicles or clumsy pedestrians and most owners don't care to replace them. Since they are completely customisable owners often just put old spare parts without connecting the rod and drives of the mirror adjustment system;

» Washers, headlights, wipers. If the seller is talking about fast emptying windshield cleaning fluid then most likely, the pump is faulty. Mentally set aside a few hundred more for repair;

» Seat adjustment system. For some reason, French cars have fragile seats. The most important to check is the driver's one. If all the seats have clear signs of wear, you are most probably seeing a car which has been used as a taxi previously;

» Scratches under the door handles, the clarity of their paint. Even the most expensive cars have a little tighter passenger door handles. Strong scratches or other minor defects under the handles, as well as a suspiciously loose handle can again suggest that this car served as a workhorse for a taxi driver;

» Smoke stains inside the cabin. Even if, you yourself are a heavy smoker like a steam locomotive, an interior stained with cigarette smoke is a good reason for bargaining. Therefore, inspect the ceiling and racks, try to find traces of tobacco smoke;

» Cleanliness of the seats. If the seats are covered with covers, check after removing them. The upholstery of the seats is not always perfectly clean, they hide dirt, cigarette burns and other unpleasant things under the fabric;

» Seat belts. Pull all the belts, they should be easily pulled out and when you jerk them out really fast they should stop. Any sign of damage to seat belts is inacceptable. Belt locks should easily lock and release the buckle.

Every time you discover something suspicious question the seller about it. An honest person has no reason to hide the facts and you are more likely to receive truthful answers. If the seller is trying to play you or cheat you, then he might try to avoid your question or lie profoundly. And the more he lies, the less you should consider purchasing the car. If some defect is hiding so carefully, will the purchase not be worth the additional financial cost?

Do not hesitate to ask questions. You are a buyer who is paying a lot of money. The safety of you and your close ones depends on the technical condition of the car you are bargaining for! So don't take any chances and do not believe in words, check and ask. Remember that no matter how tempting the offer might look, you can always find a less problematic car!

Write down all the defects on a notebook for reference. You can use this list to shoot down the price when bargaining.

AFTERWORD

In this guide, we have tried discuss all the details of buying a used car. Starting from the choice of model and concluding with the necessary steps to be taken after purchasing the car. But keep in mind that every car, even a new one is unique. There are frequent cases like for example, the new Skoda Octavia had failure of components one after the other right out of the factory, which makes car ownership extremely problematic. So what about a used vehicle? Yes, it can fail and should be checked, but there is no 100% guarantee that the transmission will not fail and engine will not end up knocking! And yet the more careful measures you take to inspect and analyse the car before you buy it, the less money you would require to invest on it.

Wishing you good luck on the road!

Your,

Allan Black